Natural Element Method for the Simulation of Structures and Processes

Natural Element Method for the Simulation of Structures and Processes

Francisco Chinesta
Serge Cescotto
Elías Cueto
Philippe Lorong

First published 2011 in Great Britain and the United States by ISTE Ltd and John Wiley & Sons, Inc.
Adapted and updated from *La méthode des éléments naturels en calcul des structures et simulation des procédés* published 2009 in France by Hermes Science/Lavoisier © LAVOISIER 2009

ISTE Ltd
27-37 St George's Road
London SW19 4EU
UK

John Wiley & Sons, Inc.
111 River Street
Hoboken, NJ 07030
USA

www.iste.co.uk

www.wiley.com

Library of Congress Cataloging-in-Publication Data

Chinesta, Francisco.
 Natural element method for the simulation of structures and processes / Francisco Chinesta.
  p. cm.
 Summary: "This book presents a recent state of the art on the foundations and applications of the meshless natural element method in computational mechanics, including structural mechanics and material forming processes involving solids and Newtonian and non-Newtonian fluids"-- Provided by publisher.
 Includes bibliographical references and index.
 ISBN 978-1-84821-220-6 (hardback)
 1. Materials--Mechanical properties--Mathematical models. 2. Numerical analysis. 3. Numbers, Natural. I. Title.
 TA404.8.C486 2011
 624.1'7015118--dc22

2010048621

British Library Cataloguing-in-Publication Data
A CIP record for this book is available from the British Library
ISBN 978-1-84821-220-6

Printed and bound in Great Britain by CPI Antony Rowe, Chippenham and Eastbourne.

Contents

"Fondation Cetim" 2008 Award

Foreword

Cetim, an 850-member strong organization, is recognized nationally and internationally due to its applicable research and contribution to the innovation of the mechanical engineering industry. In this world of globalization, with exchanges and the increasing exacerbated competition, it became apparent to us to prepare the long-term future by expanding basic research through the creation of Cetim. This was done to encourage university laboratories to develop scientific works in response to the challenges of the competitive mechanical engineering industry.

Since its foundation in 2003, Cetim has given financial support to around 10 large scientific projects, associating approximately 40 laboratories.

But once these scientific advances have been obtained, the most important thing is to make them known to the community, especially to engineers and industrial mechanics. This is why Cetim also promotes the publication of results in a monographic form, leaving a great scope for potential applications of this new knowledge, by allotting each year one or more awards to the authors of particularly remarkable monographs.

The Cetim award was thus allotted in 2008 to the research team that carried out this work and devoted itself to establishing the foundations of an original and promising method, the natural element method, in the field of the numerical simulation of structures and material forming processes.

Numerical simulation is indeed advancing with great steps and is of great importance to the mechanical engineering industry. Simulation certainly makes it possible to guarantee the speed and reliability of conception, to open the field of innovation, and especially to reduce the time and cost of conception. A survey led by the research company Aberdeen Group carried out in 2006 on mechanical manufacturers revealed that the large manufacturers who intensely used the numerical simulation of structures and processes made less than half physical prototypes than the average manufacturer, and put their products on the market 58% early, with the design cost being 50% lower.

The natural element method should make a significant contribution in response to the vast competitiveness of mechanical companies.

Michel Laroche
Chairman of Cetim

Acknowledgements

The authors would like to extend their thanks to everyone who contributed to the results shown or described in this book, especially: I. Alfaro, D. Gonzalez and M. Doblare from the University of Saragosse; J. Yvonnet from the University of Marne-la-Vallée; L.A. Illoul from Arts et Métiers ParisTech; as well as M. LI Xiang, doctorate at the University of Liège (Belgium) and the University of Technology of Dalian (People's Republic of China).

Chapter 1

Introduction

"Meshless" methods are alternative techniques to the finite element method in solving partial differential equations. While the finite element method derives an approximation based on the elements, using shape functions, the meshless methods allow us to derive an approximation at any point; thanks to the information provided by the surrounding nodes. In these approaches the concept of element is thus not used any more. Connectivity between the nodes is not defined any more by the mesh but only by the concepts of "vicinity" or "field of influence." These methods were developed with the aim of avoiding the numerical problems involved in mesh construction. These problems have been discussed in many studies; it is, for example, a question of simulation of manufacturing processes such as extrusion, injection, or setting forms by removal of matter where it is necessary to face extremely large distortions of the mesh. In other processes such as foundry, drilling, or laser welding, precisely knowing the position of the interface between the solid phase and the liquid phase is essential. In the simulation of processes such as cutting by adiabatic shearing which involves a localized deformation, possibly accompanied by the propagation of a fissure, it is necessary to carry out the simulation without the mesh being conceived influencing the direction of propagation of the shear band or the fissure. The appearance of a localized deformation requires a finer representation of the solution in certain areas of the domains, and it is thus necessary to be able to refine the mesh easily without the geometrical constraints known within the framework of finite elements (mainly in 3D) and the problems related to precise projection of the fields between the two meshes. The objective of the meshless methods is to eliminate the structure of the mesh and to build the approximation starting only from the nodes. Although structures with a geometrical character are necessary (to build node connectivity for the integration of the weak form associated with the equation to be solved and so on), these do not interfere, in general, with the quality of the solution and thus can be built independently. Even after being proposed at the end of 1970s, the "meshless" methods had to wait approximately 15 years before having a real development and an interest within the scientific community.

In the interval, little passion had been shown for them because of the numerous difficulties presented by the first techniques. The first "meshless" method seems to be the so-called smooth particle hydrodynamics (SPH) method (Lucy 1977), which was initially used to model astronomical phenomena in unbounded domains. This method, based on an approximation using the properties of the convolution product, has two disadvantages: low consistency and difficulty associated with the imposition of boundary conditions. In 1992, Nayrolles, Touzot and Villon proposed using a local approximation of least squares in a new method called the "diffuse elements method" (DEM). In 1994, Belytschko *et al.* proposed the "element-free Galerkin" (EFG) method based on the same principles as the preceding one but using "exact" derivatives of the shape functions. The method known as the "reproducing kernel particle method" (RKPM) introduced by Liu *et al.* in 1995 is an extension of the SPH method but with the reproduction of linear fields or polynomials of higher order being introduced, thanks to the correction function affecting the kernel function used in SPH method. Finally, the so-called partition of unity method introduced by Babuska in about 1996 is a general principle allowing us to enrich any function associated with a problem involving known physics, within the framework of finite elements and of meshless methods, by adding additional unknowns in the global system of equations. Thus, particular functions such as discontinuous functions and singular functions can be reproduced.

Lastly, more recently, the natural element method (NEM) rests on principles completely different from the previous ones. This method is halfway between meshless methods and the finite element method. The NEM proposes an interpolation based on the concepts of the Voronoi diagram and its natural neighbors. The Voronoi diagram associated with a cloud of nodes distributed over the domain to be studied is the Delaunay dual mesh. Thus, a mesh is being used for the construction of the interpolation. However, as the examples presented in this chapter show, the quality of interpolation produced does not depend on the form of the triangles (2D problems) or tetrahedrons (3D problems) present in the Delaunay mesh. The latter is built in a systematic way without requiring repositioning of nodes. With NEM the choice of support of shape functions is automatic and optimal in the sense that node vicinity is taken into account as much as possible to define the interpolation. With regard to the imposition of boundary conditions, for *convex domains*, it is direct and proceeds as the finite elements: the influence of internal nodes on a given domain is cancelled on the edges of the latter. The NEM cumulates the advantages of meshless methods and finite element approaches even if, with respect to the latter, a surcharge exists for the construction of the interpolation.

To extend these characteristics to the *non-convex domains* two strategies exist. The first approach based on the alpha forms makes it possible to introduce a description of the border in a very flexible way if the latter remains slightly non-convex. In the case of strongly non-convex fields, the constrained NEM (CNEM) proposes to build the interpolation on a constrained Voronoi diagram, which is the constrained Delaunay dual mesh. For the second approach, in addition to the node cloud, a valid description of the border of the field must be introduced. The Delaunay mesh is constrained with respect to this border.

The purpose of this text is to describe the technique of natural elements in its context, i.e. compared to the techniques of finite elements type, which have proved reliable for many years, but also compared to other techniques with and without meshes. Both advantages and disadvantages have been listed. This book has been written with a teaching purpose to be used by both professionals and students at Master's level. Many examples have been discussed to illustrate our remarks in order to show the potentialities of the approach. The majority of these examples will be from the framework of simulation of methods of working where the application of the meshless techniques takes all its direction owing to the great material transformations that seriously compromise the effectiveness of the techniques based on the existence of a mesh in Lagrangian formulation (or updated Lagrangian).

To better understand the context how the NEM appeared, we will revisit the main so-called meshless techniques that preceded it, for which these techniques will be described and discussed briefly and many references will be provided to allow the reader to further develop their comprehension.

1.1. SPH method

The SPH (Smooth Particle Hydrodynamics) method was introduced for solving astrophysics models. It is based on an approximation built starting from an integral of a convolution product

$$u^h(x) = \int u(y)W(x - y, h)d\Omega_y,$$ [1.1]

where the kernel function has the following properties:

- $\lim_{h \to 0} \int u(y)W(x - y, h)d\Omega_y = u(x)$;
- $\int W(x - y, h)d\Omega_y = 1$;
- W has a compact support;
- W is decreasing with distance;
- $W(x - y, h) \in \mathcal{C}^p(\mathbb{R}^n)$, $p \geq 1$.

The first property expresses that the function core tends toward the Dirac distribution, which is the limit in which equation [1.1] makes sense. The second property ensures zero-order consistency of the approximation. The third ensures the locality of the approximation and thus after discretization it will lead to a sparse linear system. The last condition allows us to obtain certain regularity in the resulting approximation.

One of the most used kernel functions is that using a Gaussian

$$W(x, h) = \frac{1}{(\pi h^2)^{n/2}} \exp\left[-\frac{x^2}{h^2}\right]; \qquad [1.2]$$

even if other kernels are also often used (splines, etc.).

In general, the support of these kernel functions is circular (in 2D) or spherical (in 3D), but it is also possible to make it ellipsoidal or even rectangular (Figure 1.1) with the introduction of tensorial products:

$$W(x - x_I) = W(x - x_I)W(y - y_I). \qquad [1.3]$$

The integral in equation [1.1] can be discretized using a nodal quadrature

$$u^h(x) = \sum_{I:x\in\Omega_I} W(x - x_I)\Delta V_I u_I, \qquad [1.4]$$

with ΔV_I the volume associated with each node. The approximation can thus be rewritten in its more usual form:

$$u^h(x) = \sum_{I:x\in\Omega_I} \Phi_I(x)u_I, \qquad [1.5]$$

where the functions Φ_I represent the shape functions of the approximation.

The fact of having a poor consistency explains why SPH approximation was especially used in the discretization of the continuous models (partial differential equations) in its strong formulation by means of a collocation scheme.

Many difficulties have been listed concerning the use of SPH approximation. These difficulties justified many works trying to circumvent these difficulties with more or less success. We can recount some of them: imposition of the essential boundary conditions (Dirichlet) due to a non-interpolant character of the approximation (equation [1.5]), i.e. the shape functions associated with the interior nodes, the support of which has a non-zero intersection with the edge of the field, are not cancelled

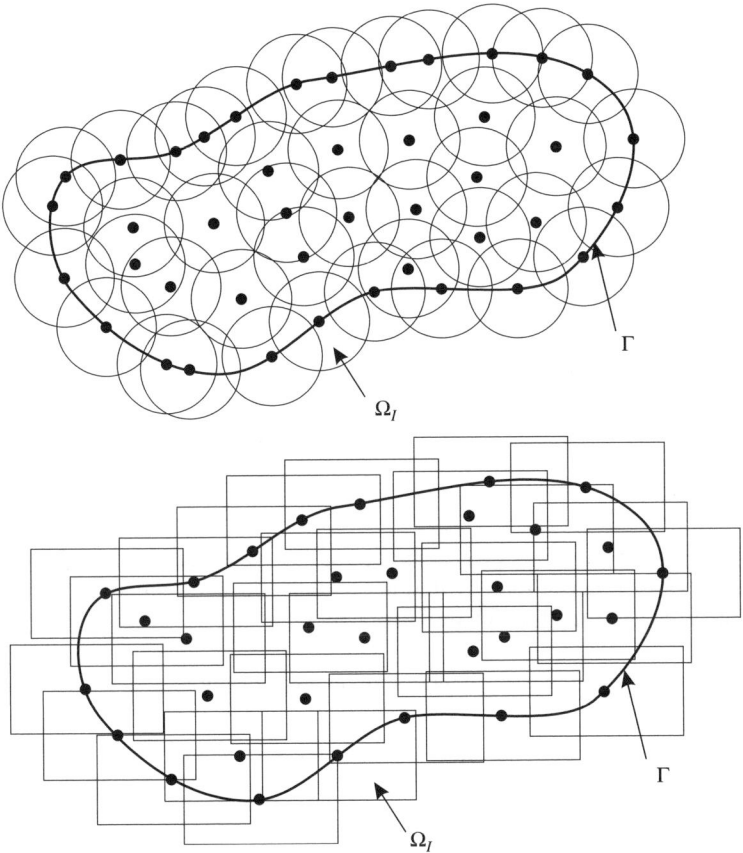

Figure 1.1. *Recovery of a 2D field*

out on the domain boundary; instabilities observed in the solids subjected to states of tensile stress that justified the introduction of the "stresses points" among other solutions; parasitic modes of deformation; and the inconsistency mentioned already.

These problems of inconsistency justified the proposal of corrections of the kernels for ensuring consistency, a step that led to RKPM.

1.2. RKPM method

To simplify the explanation we consider here a unidimensional domain as the support of the problem to be solved (all the results can be extended to the 2D or 3D case). The points in this domain will be represented by x or s.

1.2.1. *Conditions of reproduction*

The approximation $u^h(x)$ of $u(x)$ is derived from the convolution product:

$$u^h(x) = \int_\Omega w(x - s, h)u(s)\mathrm{d}\Omega, \qquad [1.6]$$

where $w(x - s, h)$ is the kernel function and h a parameter that controls the support of the approximation.

The main idea in the RKPM is to force the approximation to reproduce an unspecified function. By simplicity we will suppose that the function, which we want to reproduce exactly, is written as the sum of a polynomial part and non-polynomial part $u^e(x)$:

$$u^h(x) = a_0 + a_1 x + \cdots + a_n x^n + a_{n+1} u^e(x). \qquad [1.7]$$

In the following section, we will be discussing regarding the properties that the kernel function will have to satisfy in order to define an approximation which will be able to reproduce exactly the function in equation [1.7].

From equation [1.6], the reproduction of a constant function a_0 is written as

$$\int_\Omega w(x - s, h)a_0\mathrm{d}\Omega = a_0, \qquad [1.8]$$

which implies

$$\int_\Omega w(x - s, h)\mathrm{d}\Omega = 1, \qquad [1.9]$$

which is none other than the partition of unity.

The condition to be verified in order to reproduce a linear function $u^a(x) = a_0 + a_1 x$ is expressed in the same way by

$$\int_\Omega w(x - s, h)(a_0 + a_1 s)\mathrm{d}\Omega = a_0 + a_1 x. \qquad [1.10]$$

Using the partition of unity (equation [1.9]), equation [1.10] can be modified to the form

$$\left\{ \begin{array}{l} \int_\Omega w(x - s, h)\mathrm{d}\Omega = 1 \\ \int_\Omega w(x - s, h)s\mathrm{d}\Omega = x \end{array} \right. \qquad [1.11]$$

implying the linear consistency of the approximation. By repeating procedure, we can derive the n-order reproduction

$$\begin{cases} \int_\Omega w(x - s, h)\mathrm{d}\Omega = 1 \\ \int_\Omega w(x - s, h)s\mathrm{d}\Omega = x \\ \qquad \vdots \\ \int_\Omega w(x - s, h)s^n\mathrm{d}\Omega = x^n. \end{cases} \qquad [1.12]$$

Consequently, by rewriting the function present in equation [1.7] we get

$$\int_\Omega w(x - s, h)(a_0 + a_1 s + \cdots + a_n s^n + a_{n+1} u^e(s))\mathrm{d}\Omega =$$

$$a_0 + a_1 x + \cdots + a_n x^n + a_{n+1} u^e(x), \qquad [1.13]$$

from which we deduce

$$\begin{cases} \int_\Omega w(x - s, h)\mathrm{d}\Omega = 1 \\ \int_\Omega w(x - s, h)s\mathrm{d}\Omega = x \\ \qquad \vdots \\ \int_\Omega w(x - s, h)s^n\mathrm{d}\Omega = x^n \\ \int_\Omega w(x - s, h)u^e(s)\mathrm{d}\Omega = u^e(x). \end{cases} \qquad [1.14]$$

In the original procedure suggested by Liu *et al.* [LIU 95], only polynomial consistency of degree n was imposed. However, this procedure cannot be directly generalized to impose the reproduction of a generic non-polynomial function $u^e(x)$.

1.2.2. *Correction of the kernel*

We will represent by $u^r(x)$ the function of approximation verifying the conditions stated in the system of equations [1.14]. Normally, the kernel function is taken in the form of a cubic spline function, and consequently, the equations [1.14] are not satisfied. Liu *et al.* [LIU 95] proposed the introduction of a correction function $C(x, x - s)$ to satisfy all the conditions of reproduction. In our case, where we want to reproduce any polynomial or non-polynomial function also, we will consider the more general form $C(x, s, x - s)$, the relevance of which will be discussed later. Thus, $u^r(x)$ can be expressed by [TRU 05]:

$$u^r(x) = \int_\Omega C(x, s, x - s)w(x - s, h)u(s)\mathrm{d}\Omega, \qquad [1.15]$$

where $C(x, s, x - s)$ is sought in the form

$$C(x, s, x - s) = \mathbf{H}^T(x, s, x - s)\mathbf{b}(x), \qquad [1.16]$$

where $\mathbf{H}^T(x, s, x - s)$ represents the vector containing the functions considered in the approximation basis and $\mathbf{b}(x)$ is another vector whose components are unknown functions that will be determined to verify the conditions of reproduction. Thus, equation [1.14] can be rewritten as:

$$\begin{cases} \int_\Omega \mathbf{H}^T(x, s, x - s)\mathbf{b}(x)w(x - s, h)\mathrm{d}\Omega = 1 \\ \int_\Omega \mathbf{H}^T(x, s, x - s)\mathbf{b}(x)w(x - s, h)s\mathrm{d}\Omega = x \\ \qquad \vdots \\ \int_\Omega \mathbf{H}^T(x, s, x - s)\mathbf{b}(x)w(x - s, h)s^n\mathrm{d}\Omega = x^n \\ \int_\Omega \mathbf{H}^T(x, s, x - s)\mathbf{b}(x)w(x - s, h)u^e(s)\mathrm{d}\Omega = u^e(x). \end{cases} \qquad [1.17]$$

In fact, the conditions of reproduction must be imposed in discrete form. To do so, we will consider N points (also called nodes) allowing us to calculate the discrete form of equation [1.17]:

$$\begin{cases} \sum_{i=1}^N \mathbf{H}^T(x, x_i, x - x_i)\mathbf{b}(x)w(x - x_i, h)\Delta x_i = 1 \\ \sum_{i=1}^N \mathbf{H}^T(x, x_i, x - x_i)\mathbf{b}(x)w(x - x_i, h)x_i\Delta x_i = x \\ \qquad \vdots \\ \sum_{i=1}^N \mathbf{H}^T(x, x_i, x - x_i)\mathbf{b}(x)w(x - x_i, h)x_i^n\Delta x_i = x^n \\ \sum_{i=1}^N \mathbf{H}^T(x, x_i, x - x_i)\mathbf{b}(x)w(x - x_i, h)u^e(x_i)\Delta x_i = u^e(x), \end{cases} \qquad [1.18]$$

whose matrix form is

$$\left[\sum_{i=1}^N \mathbf{R}(x_i)\mathbf{H}^T(x, x_i, x - x_i)w(x - x_i, h)\Delta x_i \right] \mathbf{b}(x) = \mathbf{R}(x), \qquad [1.19]$$

where $\mathbf{R}(x)$ represents the vector of reproduction,

$$\mathbf{R}^T(x) = [1, x, \ldots, x^n, u^e(x)]. \qquad [1.20]$$

Equation [1.19] enables us to calculate the vector $\mathbf{b}(x)$,

$$\mathbf{b}(x) = \mathbf{M}(x)^{-1}\mathbf{R}(x), \qquad [1.21]$$

where the so-called moment matrix $\mathbf{M}(x)$ is defined by:

$$\mathbf{M}(x) = \sum_{i=1}^{N} \mathbf{R}(x_i)\mathbf{H}^T(x, x_i, x - x_i)w(x - x_i, h)\Delta x_i. \qquad [1.22]$$

This matrix differs slightly from that obtained in [LIU 95].

1.2.3. *Discrete form of the approximation*

The discrete form $u^r(x)$ of $u^h(x)$ is obtained from equations [1.15], [1.16] and [1.21]:

$$u^r(x) \cong \sum_{i=1}^{N} \mathbf{H}^T(x, x_i, x - x_i)\mathbf{M}(x)^{-1}\mathbf{R}(x)w(x - x_i, h)u(x_i)\Delta x_i$$

$$= \sum_{i=1}^{N} \psi_i(x)u_i, \qquad [1.23]$$

where ψ_i is the shape function associated with the enriched RKPM approximation:

$$\psi_i(x) = \mathbf{H}^T(x, x_i, x - x_i)\mathbf{M}(x)^{-1}\mathbf{R}(x)w(x - x_i, h)\Delta x_i. \qquad [1.24]$$

As in the most standard version of the RKPM, we take $\Delta x_i = 1$. Although various quadratures exist, the choice of the quadrature does not affect the precision of the constructed approximation.

If this method allows us to overcome a certain number of difficulties present in the SPH method, the difficulty related to the imposition of the essential boundary conditions remains untouched. The gain on the side of consistency allowed the use of RKPM approximations within the framework of discretizations of the weak (often variational) formulations of partial differential equations. Although this possibility of working on weak formulations seems to be a positive point, it hides a new difficulty associated with the integration of the weak forms. Integration requires a decomposition of the domain and use of a suitable quadrature formula. Moreover, in certain cases, the non-polynomial character of the resulting shape functions makes this integration delicate. This subject has also motivated many studies.

It is possible to prove that the RKPM, which we have just described briefly, is completely equivalent to another family of methods. These methods will be described later and are based on the use of approximations making use of the moving least squares (*MLS*).

1.3. MLS based approximations

We now will consider the approximation:

$$\mathbf{u}^h(\mathbf{x}) = \mathbf{p}^T(\mathbf{x})\mathbf{a}(\mathbf{x}), \qquad [1.25]$$

with $\mathbf{p}^T(\mathbf{x})$ a polynomial base. For example, $\mathbf{p}^T(\mathbf{x}) = [1, x, y, xy]$ and $\mathbf{p}^T(\mathbf{x}) = [1, x, y, xy, x^2, y^2]$, respectively, represent a linear and quadratic base in the 2D case and $\mathbf{a}(\mathbf{x})$ represents a vector with unknown coefficients. To determine $\mathbf{a}(\mathbf{x})$, we will define the functional calculus J which will have to be minimized with respect to $\mathbf{a}(\mathbf{x})$ [NAY 92]:

$$J = \frac{1}{2}\sum_{i=1}^{n} w_i(\mathbf{x})\left[\mathbf{p}^T(\mathbf{x}_i)\mathbf{a}(\mathbf{x}) - u_i\right]^2, \qquad [1.26]$$

where u_i are the nodal unknown associated with nodes $\mathbf{x}_i$ neighboring points of $\mathbf{x}$ and $w_i(\mathbf{x})$ is a weight function whose value decreases with the distance between $\mathbf{x}_i$ and $\mathbf{x}$ (refer to [BEL 98a] to understand the main properties of this function as well as the most used weight functions). The minimization of J with respect to the coefficients $a_j(\mathbf{x})$ led to:

$$\frac{\partial J}{\partial a_j(\mathbf{x})} = \sum_{k=1}^{n} a_k \left[\sum_{i=1}^{n} w_i(\mathbf{x})p_j(\mathbf{x}_i)p_k(\mathbf{x}_i)\right] - \sum_{i=1}^{n} w_i(\mathbf{x})p_j(\mathbf{x}_i)u_i = 0, \qquad [1.27]$$

which led to the linear system:

$$\mathbf{A}(\mathbf{x})\mathbf{a}(\mathbf{x}) = \mathbf{B}(\mathbf{x})\mathbf{u}, \qquad [1.28]$$

where the matrices $\mathbf{A}(\mathbf{x})$ and $\mathbf{B}(\mathbf{x})$ are defined by:

$$A_{jk}(\mathbf{x}) = \sum_{i=1}^{n} w_i(\mathbf{x})p_j(\mathbf{x}_i)p_k(\mathbf{x}_i), \qquad [1.29]$$

$$B_{ij}(\mathbf{x}) = w_i(\mathbf{x})p_j(\mathbf{x}_i). \qquad [1.30]$$

While replacing $\mathbf{a}(\mathbf{x})$ in equation [1.25], we obtain:

$$u^h(\mathbf{x}) = \mathbf{p}^T(\mathbf{x})\mathbf{A}^{-1}(\mathbf{x})\mathbf{B}(\mathbf{x})\mathbf{u}, \qquad [1.31]$$

from which we can identify the shape functions of the approximation:

$$\psi^T(\mathbf{x}) = \mathbf{p}^T(\mathbf{x})\mathbf{A}^{-1}(\mathbf{x})\mathbf{B}(\mathbf{x}).$$ [1.32]

The only difference between the diffuse approximation and approximation used in the technique known as *EFG* resides in the evaluation of derivatives of the shape functions. In the first technique, only the functions of the approximation base contained in vector $\mathbf{p}^T(\mathbf{x})$ in equation [1.32] are derived, while in the second all the terms depending on $\mathbf{x}$ are there. It amounts to the saying that not only are the derivatives of the functions contained in the base of approximation considered but also the derivatives of the coefficients $a_j(\mathbf{x})$ present in the approximation.

1.4. Final note

Although we have summarized only the most known methods here, there exist various techniques which can be listed as pertaining to the family of meshless techniques. Since many works and articles in specialized papers have been devoted to these techniques, we do not want to expand on them further. We will quote simply some other techniques: generalized finite differences, the "h-p clouds;" finite sphere methods; and the methods based on the partition of unity such as the generalized finite elements or even the extended finite elements (X-FEM) methods which, although based on the finite element method, succeed in freeing themselves from certain difficulties related to the management of interfaces or discontinuities, fixed or moving without calling on the traditional remeshing techniques.

Chapter 2

Basics of the Natural Element Method

2.1. Introduction

For the past 15 years, and especially after the study by Nayroles, Touzot and Villon [NAY 92] regarding the diffuse finite elements, the meshless methods attracted much interest in the *computational mechanics community*. For more information, refer to the recent study by [FER 04] and [BAB 03]. The main advantage of these methods is that they can describe large transformations of the material domain without being restricted by mesh distortion, as opposed to the finite element methods (FEMs), while avoiding the stage associated with the mesh of the model. Moreover, from the point of view of the simulation of free-surface problems, the propagation of fissures, or all those integral to the internal variables, these methods seem promising because the evolution of these internal variables can be calculated on the nodal trajectories, without requiring projection phases, with all their problems.

In spite of their advantage, the majority of the meshless methods have certain difficulties, for example, the imposition of essential boundary conditions in non-convex domains or the precise numerical integration of the associated variational formulations.

The natural neighbor Galerkin methods [SUK 01b], also known as the natural FEM [TRA 94], have specific properties with regard to other meshless methods. This is the case with the interpolating nature of their shape functions, which, with the property relating to the cancellation of the shape functions associated with the internal nodes on the convex edges, allows simple and direct imposition (as in FEM) of the essential boundary conditions.

The natural neighbor Galerkin methods use an interpolation based on the natural neighbors, in place of the polynomial compact support function characteristic of the finite elements, the moving least squares type, or the other meshless methods based on the method of particles (smooth particle hydrodynamics, reproducing kernel particle method, and so on). This interpolation is presented in section 2.2, and the approach to the problems related to the imposition of the boundary conditions is discussed in section 2.3. A delicate question specific to the mixed formulations, and which is found with the meshless approaches, is the verification of the stability conditions to ensure the convergence of the numerical solution. These conditions are known as inf-sup or Ladyzhenskaya-Babuska-Brezzi (LBB) conditions, and the case of the natural neighbor interpolation is discussed in section 2.4. Finally, the construction of high degree natural neighbor interpolations (allowing us to have arbitrary consistency and smoothness) is described in section 2.5.

2.2. Natural neighbor Galerkin methods

In this section, we describe the interpolation, which is used throughout this study for the discretization of the variational formulations in a Galerkin framework [SUK 01b]. Before defining the interpolation, we need to know the geometrical concepts associated with the Voronoi diagram and the Delaunay triangulation of a node cloud.

2.2.1. *Interpolation of natural neighbors*

The concept of the Voronoi diagram was originally introduced by mathematicians Dirichlet [DIR 50] and Voronoi [VOR 08] and later applied in many scientific disciplines. The Voronoi diagram of a node set in $\mathbb{R}^n$ divides the space of dimension n into areas T_I, each of which is associated with a node n_I, such that any point inside one of these areas is nearer to the node defining the cell than any other node.

The Voronoi diagram is unique for a given set of nodes. It carries out a partition of space and can be extended to any dimension. The Voronoi diagram is formally given by

$$T_I = \{\boldsymbol{x} \in \mathbb{R}^n : d(\boldsymbol{x}, \boldsymbol{x}_I) < d(\boldsymbol{x}, \boldsymbol{x}_J) \, \forall J \neq I\}, \qquad [2.1]$$

where T_I is the Voronoi cell associated with a node n_I, $\boldsymbol{x}$ is the position of a generic point x, $\boldsymbol{x}_I$ defines the coordinates of the node n_I, and $d(\boldsymbol{x}, \boldsymbol{x}_J)$ is the distance (Euclidian norm) between the node n_J and the point x.

In Figure 2.1, on the right, we illustrate a degenerated situation, with two possible triangulations: this situation is found when four points are located on the same circumscribed circle.

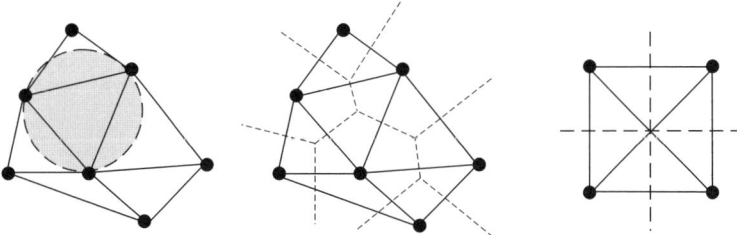

Figure 2.1. *Delaunay triangulation and Voronoi diagram*

The Voronoi cells T_I are strictly convex polygons (polyhedral in three-dimensional (3D)) and not limited to the nodes present on the convex envelope of the domain. By connecting the nodes sharing a common side of the Voronoi cell, we obtain the Delaunay triangulation (introduced by Voronoi [VOR 08] and developed by Delaunay [DEL 34]), which is the dual of the Voronoi diagram. The circumscribed circles with the Delaunay triangles do not contain any nodes (Figure 2.1). Finally, the vertices of Voronoi cells are the orthocenters of the Delaunay triangles and the centers of the circumscribed circles with these triangles. These definitions are widely used in 3D (thus, the triangles are tetrahedrons, the circles are spheres, and the Voronoi polygons are polyhedrons).

The natural neighbors of a node are the nodes whose Voronoi cell shares an edge with the one of the considered nodes or which are connected to the node by an edge of the Delaunay triangle (side of a tetrahedron in 3D). We can note that in all these cases, even when the position of the nodes is irregular, that is, the distance between nodes is significant in certain areas or the nodal distribution is strongly anisotropic, the Delaunay triangulation is always the best possible triangulation (in 2D) from a finite element point of view. This triangulation maximizes the minimum interior angle of the traingles. This is not generally true in 3D.

The simplest natural neighbor-based interpolation scheme is based on the nearest neighbor (known as Thiessen's interpolation [THI 11]). In this diagram, the nodal value is associated with the whole cell, hence the resulting continuity $\mathcal{C}^{-1}$. Evidently, this interpolation is not adapted for the discretization of the second-order partial differential equations, but it will be used for the construction of the mixed velocity-pressure approximations [GON 04b], which will then be applied to the discretization of the problems of incompressible solids or fluids.

The most popular type of natural neighbor interpolant is accredited to Sibson [SIB 81]. Before leading to its description, we must define the second-order Voronoi cell, containing the points with the first closest node as the n_I, and the second nearest node as the n_J:

$$T_{IJ} = \{\boldsymbol{x} \in \mathbb{R}^n : d(\boldsymbol{x}, \boldsymbol{x}_I) < d(\boldsymbol{x}, \boldsymbol{x}_J) < d(\boldsymbol{x}, \boldsymbol{x}_K) \ \forall K \neq J; \ K \neq I\}. \quad [2.2]$$

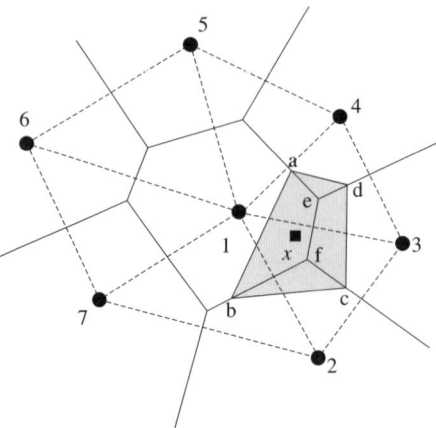

Figure 2.2. *Definition of the natural coordinates of a point* x

If a point is added (a point where we try to determine the value of the various shape functions — typically a point of integration), the Voronoi diagram will be modified (Figure 2.2). In [SIB 80], the natural neighbor coordinates of point x with respect to its neighbor I are defined as the ratio between the area (volume in 3D) of the cell T_I that is transferred to T_x and the area associated with T_x. In other words, if $\kappa(x)$ and $\kappa_I(x)$ represent Lebesgue measurements T_x and T_{xI}, respectively, the natural neighbor coordinate x compared to node I is defined as

$$\phi_I^{\mathrm{sib}}(x) = \frac{\kappa_I(x)}{\kappa(x)}. \tag{2.3}$$

The resulting shape function depends on the relative position of the nodes. Figure 2.3 shows an example of a node that is surrounded by eight other nodes.

The resulting shape functions have remarkable properties (see [SUK 98b] or [CUE 03] for precise proof):

– The shape functions C^1 are almost everywhere, except at the nodes where they are simply continuous (see Figure 2.3).

– The resulting approximation has a linear consistency [SUK 98b].

– The partition of unity, which is derived from the preceding property, makes it possible to enrich the approximation by a known solution or to increase the order of consistency [BAB 96].

Recently, [HIY 99] generalized the form of the natural interpolant known as the Laplace interpolation, which is an interpolant closer to the earlier one with a clear advantage with regards to calculation time. It uses geometrical entities of a lower

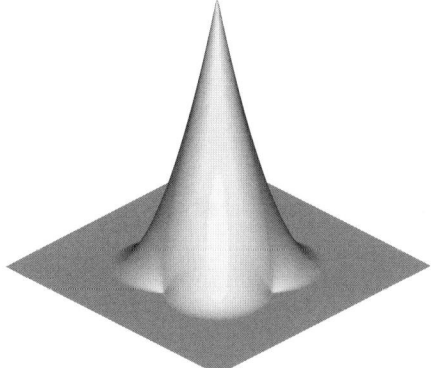

Figure 2.3. *Sibsonian shape function (courtesy of N. Sukumar)*

dimension than the dimension of the space considered. Thus, we write $t_{IJ} = \{x \in T_I \cap T_J, J \neq I\}$ (note that t_{IJ} can be the empty set), and we could define the value as

$$\alpha_J(x) = \frac{|t_{IJ}|}{d(x, x_J)}.$$ [2.4]

By referring to Figure 2.4, the value of the shape function associated with node 4 at point x is given by

$$\phi_4^{ns}(x) = \frac{\alpha_4(x)}{\sum_{J=1}^{n} \alpha_J(x)} = \frac{s_4(x)/h_4(x)}{\sum_{J=1}^{n} [s_J(x)/h_J(x)]},$$ [2.5]

where s_J represents the length of the edge common to the Voronoi cells associated with the node J and point x and n represents the number of natural neighbors of the point considered x.

The derivative of the Laplace shape functions is not defined on the edges of the Delaunay triangles contained within its support ([SUK 01b]).

In this study, we have considered only the Sibsonian interpolant. In the framework of 2D or 3D simulations, the approximation of the unknown domain (scalar, vectorial, or tensorial) is written in the form:

$$\mathbf{u}^h(x) = \sum_{I=1}^{n} \phi_I(x) \mathbf{u}_I,$$ [2.6]

where $\mathbf{u}_I$ is the vector of the nodal variables and n is the number of natural neighbors of point x. Thus, the resulting approximation is C^0.

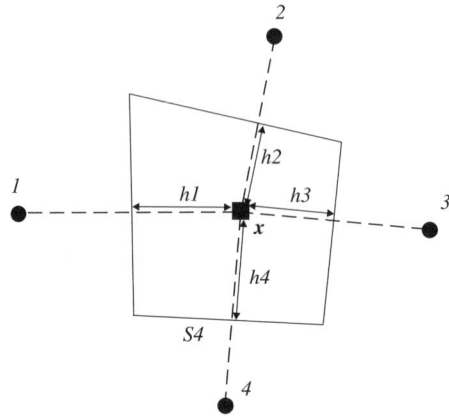

Figure 2.4. *Definition of the Laplace shape functions*

2.2.2. *Discretization*

With the earlier defined interpolation diagram, the application of the Galerkin method enables us to derive a linear system of equations.

Thus giving the generic problem,

$$\mathcal{L}(u) = f, \tag{2.7}$$

where $\mathcal{L}$ is a differential operator, u is the unknown field, and f is the source term. The associated weak formulation (known as the weighted residuals) of the preceding problem is written in the form:

$$\int_{\Omega} u^{\star} \cdot \mathcal{L}(u)\mathrm{d}\Omega = \int_{\Omega} u^{\star} f \mathrm{d}\Omega \quad \forall u^{\star}. \tag{2.8}$$

The discretization consists of approximating the unknown function and the test function by

$$\mathbf{u}^{h}(\boldsymbol{x}) = \sum_{I=1}^{n} \phi_I(\boldsymbol{x})\mathbf{u}_{\mathrm{I}}, \tag{2.9}$$

$$\mathbf{u}^{\star h}(\boldsymbol{x}) = \sum_{I=1}^{n} \phi_I(\boldsymbol{x})\mathbf{u}_{\mathrm{I}}^{\star}. \tag{2.10}$$

We continue in the same vein as for the finite elements. The fact of using the same interpolant for the unknown functions and the test functions gives rise to the (Bubnov-)Galerkin method. It is concluded that this type of interpolation could extend to more general formulations where the approximations of the unknown field and weight functions could be done using different shape functions (Petrov–Galerkin methods or formulations).

2.2.3. *Properties of the interpolant based on natural neighbors*

In contrast to other meshless approximations, the approximation that has just been defined is strictly interpolating, that is, the built approximation passes by the points that define it. This property can be written as

$$\phi_I(\boldsymbol{x}_J) = \delta_{IJ}, \qquad [2.11]$$

with δ_{IJ} the Kronecker delta function.

As is already discussed, the interpolant built has a linear consistency. The proof derives from the partition of unity (direct consequence of the construction of the interpolant) and from the property known as the *local coordinate property* [SIB 81]:

$$\sum_{I=1}^{n} \phi_I(\boldsymbol{x})\boldsymbol{x}_I = \boldsymbol{x}. \qquad [2.12]$$

Another particularly interesting aspect of the natural element method (NEM) shape functions is that they do not depend on the subjacent (Delaunay) mesh, but rather on the spatial distribution of the nodes. To illustrate this aspect, Figure 2.5 gives, for a fixed nodal distribution, and at a given point $\boldsymbol{x}$, the value of the shape function associated with each node (*full, red discs*). For NEM interpolation, the neighbors of $\boldsymbol{x}$ (*nodes of the cloud for which the associated shape functions are not zero*) are the nodes closest to $\boldsymbol{x}$: nodes j, h, k, and g (Figure 2.5). For this interpolation, the closer a node is to the place of evaluation, the bigger its influence (the *associated function value*). We note that the shape function values associated with the neighbors of $\boldsymbol{x}$ are almost identical because $\boldsymbol{x}$ is appreciably at the same distance from the latter. This is not verified for the FEM, whether we consider a strongly distorted mesh (*top mesh*) or the Delaunay mesh which minimizes this distortion (*bottom mesh*). For the top mesh, the neighbors of $\boldsymbol{x}$ are i, e, and f, which are not the nodes closest to $\boldsymbol{x}$. For the bottom mesh, the neighbors are j and g (k is also a neighbor, but the value of the associated shape function is practically zero), whereas k and h have a similar distance to $\boldsymbol{x}$.

Another remarkable property of the natural neighbor interpolation is its capacity to reproduce piece-wise linear functions on the edges of the domain if the latter is convex. We will again take the proof given by Sukumar [SUK 98b, SUK 98a], his extension to the case 3D is direct.

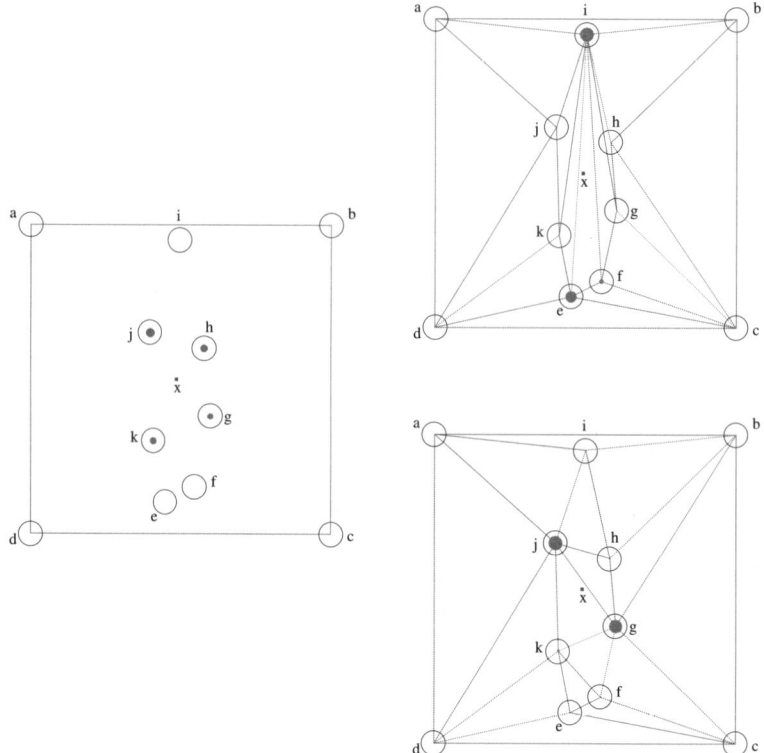

Figure 2.5. *NEM Sibson shape functions (left) and linear FEM (right) for two different meshes*

We consider two nodes placed on the edge of a convex field and a reference system ξ (see Figure 2.6). It is presumed for reasons of simplicity that point x has only three neighbors: 1, 2, and 3. The shape function, according to equation [2.3], can be calculated as follows:

$$\phi_I(\xi) = \frac{A_I(\xi)}{A(\xi)} \quad (I = 1, 2, 3),$$ [2.13]

where $A(\xi) = \sum_{J=1}^{3} A_J(\xi)$.

It is simple to note that the Voronoi cells associated with the nodes placed on the convex boundary are unbounded. The various areas are defined by

$$A_1(\xi) = \lim_{L \to \infty} L\frac{1-\xi}{2} + \delta_1, \qquad A_2(\xi) = \lim_{L \to \infty} L\frac{\xi}{2} + \delta_2, \qquad A_3 = \delta_3, \quad [2.14]$$

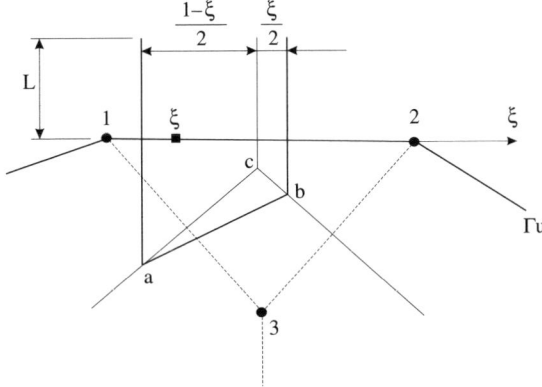

Figure 2.6. *Interpolant on the convex edge* Γ_u

where δ_I represents a finite area. From equation [2.13], we obtain

$$\phi_1(\xi) = \lim_{L \to \infty} \frac{L(1 - \xi) + 2\delta_1}{L + 2\delta_1 + 2\delta_2 + 2\delta_3},$$

$$\phi_2(\xi) = \lim_{L \to \infty} \frac{L\xi + 2\delta_2}{L + 2\delta_1 + 2\delta_2 + 2\delta_3}, \qquad [2.15]$$

$$\phi_3(\xi) = \lim_{L \to \infty} \frac{2\delta_3}{L + 2\delta_1 + 2\delta_2 + 2\delta_3}.$$

By taking the limit indicated,

$$\phi_1(\xi) = 1 - \xi, \qquad \phi_2(\xi) = \xi, \qquad \phi_3(\xi) = 0. \qquad [2.16]$$

It can be immediately noted that the contribution of the internal nodes vanish; hence, the linear character of the resulting interpolation on the edge is achieved (Figure 2.7). However, as soon as a non-convex boundary is considered, the preceding result is no longer valid. Because the areas of the nodes placed on the boundary are finite, the influence of the internal nodes is no longer negligible. Sukumar [SUK 98b] noted errors of about 2% for non-uniform distributions of nodes, finer near the edges.

Thus, as we discuss in the following section, some conditions on the node density on the boundary as on the proximity of the internal nodes at the edges have to be respected to guarantee the linearity of the approximation: the interior points should not intervene in the interpolation throughout the boundary, and the nodes on the boundary should not intervene in the interpolation of other non-visible points (by considering the boundary as opaque) [BEL 98a].

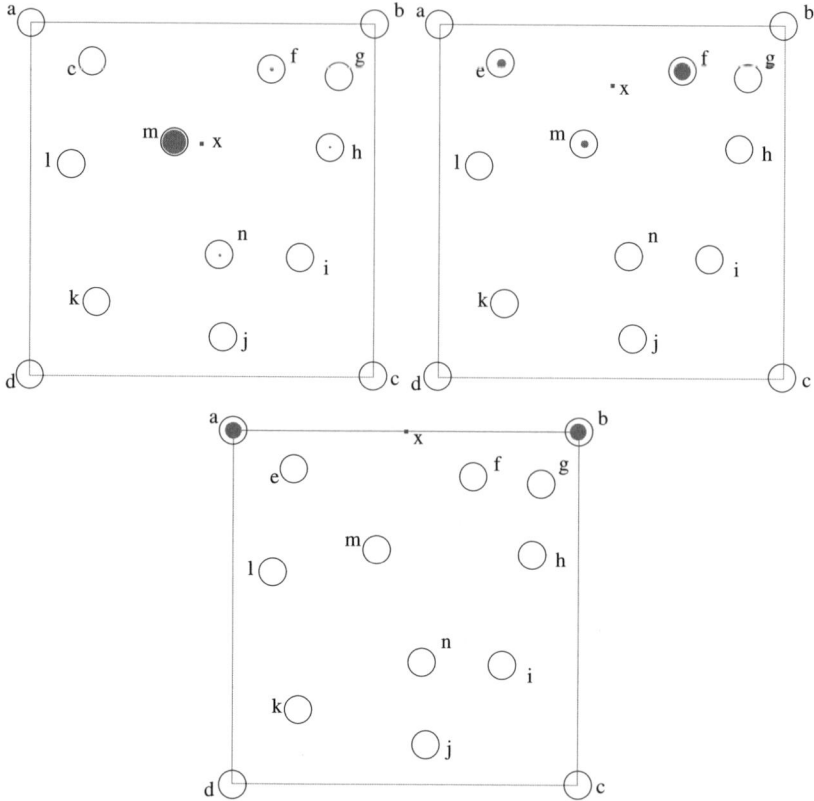

Figure 2.7. *NEM Sibson shape function close to a convex edge*

2.3. Exact imposition of the essential boundary conditions

The problems related to the imposition of the boundary conditions are a typical question of meshless methods. Some researchers who have dealt with this problem are [BEL 98a, KRO 96, GOS 96, SUK 98b]. Various techniques have been proposed: Lagrange multipliers, penalty methods, modified variational formulations, coupling with finite elements, and so on. Despite all these efforts, the problem has not been completely solved within the framework of the majority of meshless methods. The leading cause of this failure is the impossibility of making, on any type of boundary, the shape functions associated with the interior nodes vanish.

Two approaches have been proposed within the framework of the natural neighbor Galerkin method. The first approach is based on the introduction of the alpha shapes into the construction of interpolant [CUE 00]. In section 2.3.1, we briefly introduce the main ideas related to the alpha shapes and their usefulness, both for the imposition of the boundary conditions and for the automatic extraction of the boundary of the domain, which is described exclusively by the node cloud. This last property allows the application of this technique in the simulation of problems in great transformations, with the possibility of a very simple processing of free surfaces or evolving discontinuities.

The second approach is based on the use of the constrained Voronoi diagram to define the NEM interpolation, which is known as constrained natural element method (CNEM) [YVO 04a, YVO 04b, ILL 08]. The Voronoi diagram, or more precisely its dual Delaunay tetrahedralization, is then constrained to respect the boundary of the domain. This approach thus requires, in addition to the node cloud spread on the domain, a description of the boundary of the latter, which must be adapted over the course of simulation in the event of great transformations. This description of the boundary allows a strict application of the visibility criterion, specified further, and takes into account greatly non-convex domains. The CNEM approach is described in section 2.3.2.

2.3.1. *Introduction to alpha shapes*

In the case of the great transformations, and mainly in the 3D cases, the description of the evolution of the boundary of the domain becomes a delicate task: various tests must be carried out in order to correctly describe the interaction between various fronts of the flowing matter, the formation of holes, and so on. In earlier works [CUE 00, CUE 02, MAR 03], we proposed a different approach to extract the shape from a domain at each time step, without any extra information.

The concept of shape does not have a formal definition in geometry, contrary to other more familiar concepts such as diameter, curvature . . . . Edelsbrunner and Mücke [EDE 94] have parametrized the set of all the shapes related to a point cloud. For this, they introduced the concept of the alpha shape, which is associated with the level of detail with which we wish to represent the domain described by the point cloud.

An alpha shape is a polytope, not necessarily convex or connected, on which we can establish a triangulation which is a sub-set of the Delaunay triangulation of the point cloud. Thus, the criterion of the empty circumcircle will be satisfied. Let N be the point cloud in $\mathbb{R}^3$ and α be a real number, with $0 \leq \alpha < \infty$. A k-simplex σ_T with $0 \leq k \leq 3$ is defined as the convex hull of a sub-set $T \subseteq N$ of size $|T| = k + 1$. Let b be an α-ball, that is, an open ball with a radius of α. A k-simplex σ_T is α-exposed if there exists an empty α-ball b verifying $T = \partial b \bigcap N$, where ∂ indicates the boundary of the ball. In other words, a k-simplex is α-exposed if an α-ball passing by their points of definition does not contain other points of the set N.

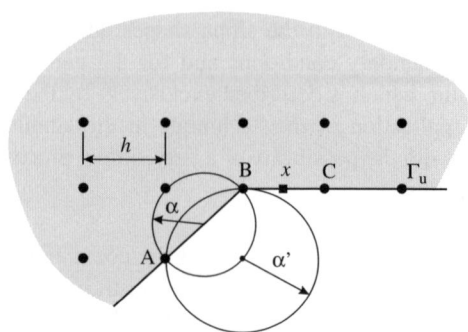

Figure 2.8. *Neighboring in the context of the α-complexes*

From this, we can define the set $F_{k,\alpha}$ as a set of α-exposed k-simplices for a set of given points N. This makes it possible to define the α shape of the set N as the polytope whose boundary is formed by the triangles belonging to $F_{2,\alpha}$, the edges in $F_{1,\alpha}$ and the nodes in $F_{0,\alpha}$.

A 3D complex is a collection $\mathcal{C}$ of closed k-simplices ($0 \leq k \leq 3$), which verifies the following:

– If $\sigma_T \in \mathcal{C}$, then $\sigma_{T'} \in \mathcal{C}$ for each $T' \subseteq T$.

– The intersection of two simplices belonging to $\mathcal{C}$ is empty, or it is limited to a face of two simplices.

Each k-simplex σ_T is included in the Delaunay triangulation $\mathcal{D}$, defined an open ball b_T whose boundary ∂b_T passes on $k + 1$ points which define the simplex. If ϱ_T is the radius of this sphere, then the $G_{k,\alpha}$ family is composed of all the k-simplices $\sigma_T \in \mathcal{D}$ whose ball b_T is empty with $\varrho_T < \alpha$. The $G_{k,\alpha}$ family does not necessarily constitute a complex. Thus, Edelsbrunner and Mücke [EDE 94] define the α-complexes $\mathcal{C}_\alpha$, as the complexes whose k-simplices either are contained in $G_{k,\alpha}$, or limit $(k + 1)$-simplices of $\mathcal{C}_\alpha$. If volume within $\mathcal{C}_\alpha$ is defined, $|\mathcal{C}_\alpha|$, as the union of all the simplices belonging to $\mathcal{C}_\alpha$, a relationship between the shape and the complexes can be established:

$$\mathcal{S}_\alpha = |\mathcal{C}_\alpha|, \quad \forall \alpha \quad 0 \leq \alpha < \infty. \tag{2.17}$$

For reasons of simplicity, we consider a 2D case with a regular distribution of N points and a non-convex boundary (Figure 2.8). With h being the distance between the nodes, the smallest value of α that reproduces the geometry is $\alpha = \frac{\sqrt{2}}{2}h$. If we consider a point x on the edge Γ_u, for an exact imposition of the essential boundary conditions, this point must have an unbounded associated cell. For this, node A, which in the Delaunay triangulation leads to a second-order Voronoi cell associated with x

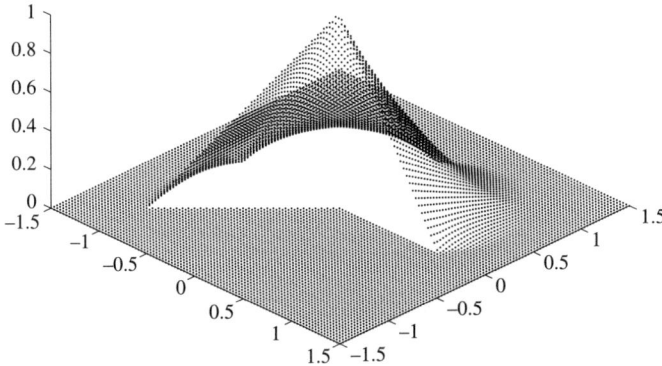

Figure 2.9. *Shape functions associated with the node B of Figure 2.8*

to have a finite area, should not be neighbor of x. The worst-case scenario takes place when x moves toward B. In this case, the α-ball that causes the union of the points A, B, and x with $\mathcal{C}_\alpha(N \bigcup x)$ has a radius $\alpha' = h$ (this is valid only if the angle is higher than 90 degrees). The shape function that results from it is shown in Figure 2.9.

We can conclude that the construction of the Sibson or Laplace interpolant on a suitable alpha shape (in order to correctly describe the expected geometry) also allows the exact imposition of the essential boundary conditions.

2.3.2. *CNEM approaches*

For the NEM, the shape function calculation is based on the Voronoi diagram, whereas for the CNEM, it is based on the Voronoi diagram *constrained* by the boundary of the domain.

The definition of the constrained Voronoi diagram is based on the criterion of visibility [ORG 96]: a node q is visible from another node p, and reciprocally, if the segment of straight line $[p, q]$ connecting them does not go through the domain boundary Γ and is not outside the domain.

In Figure 2.10, segments $[a, b]$, $[a, c]$, and $[a, d]$ go through Γ. Nodes b, c, and d are thus not visible by a and reciprocally. In addition, segment $[e, c]$ is outside the domain, and thus, the nodes e and c are not visible to each other.

The constrained Voronoi cell associated with a node p, $\mathrm{Vor}_S^c(p)$, which can be found in [BAR 93], is defined as follows:

$$\mathrm{Vor}_s^c(p) = \{x \in \mathbb{R}^n : d(x, p) \leq d(x, q), \ \forall q \in S, \ q \neq p, \ x\,p \text{ visible}, \ x\,q \text{ visible}\}. \tag{2.18}$$

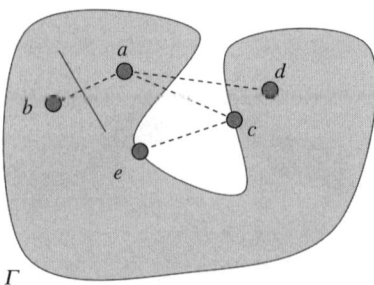

Figure 2.10. *Criterion of visibility*

This definition restricts the constrained Voronoi cell associated with a given node, to its subset within the domain. To calculate the shape functions thereafter, this area should not be *truncated* by the domain boundary.

The following is the definition of the constrained Voronoi area associated with a node where the area is not truncated by the boundary:

$$\text{Vor}_s^c(p) = \{x \in \mathbb{R}^n : d(x,p) \leq d(x,q), \ \forall q \in S, \ q \neq p \wedge p\,q \ \text{visible}\}. \qquad [2.19]$$

Figure 2.11 illustrates the non-constrained Voronoi area associated with a node p on the left and constrained Voronoi area associated with the same node on the right. The top-right cell corresponds to definition [2.18], and the bottom-right cell corresponds to definition [2.19]. For the constrained diagram resulting from definition [2.19], only the application to the nodes of the visibility criterion is used to define the cell of the node p. The nodes k, l are not visible by the node p because the segments $[p, l]$ and $[p, k]$ go through the domain boundary.

Note that the Voronoi diagram of a node cloud is the Voronoi diagram constrained by the *convex hull* of this node cloud. For a non-convex domain, the convex envelope of the nodes is no longer its boundary; hence, there is a difference between the two diagrams for such domains.

The constrained Voronoi diagram is the dual of the constrained Delaunay triangulation [SHE 98]. For this triangulation, the circumscribed circle (*sphere, hypersphere*) in each triangle (*tetrahedron, simplex*) does not contain any *interior* nodes. In 2D, the constrained Delaunay triangulation always exists either for convex or non-convex fields. This is not the case in 3D for certain non-convex domains, for example, the Schönhardt polyhedron [SCH 28] (Figure 2.12). A solution in these cases consists of adding additional nodes on the domain boundary. An algorithm by Si and Gärtner [SI 05] can be used to generate the constrained Delaunay tetrahedralization.

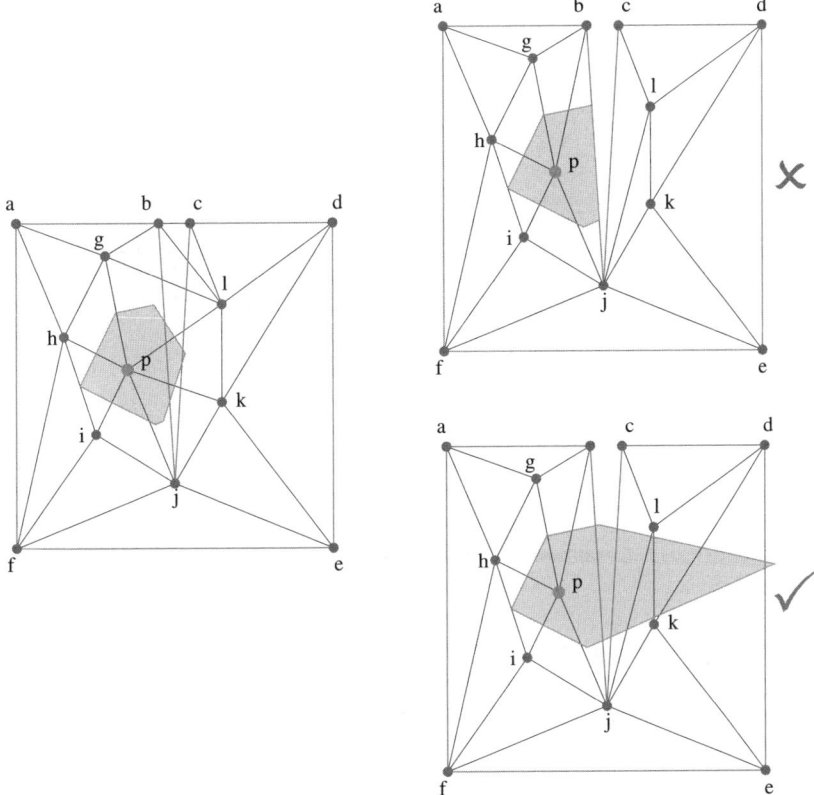

Figure 2.11. *Voronoi area associated with a node,*
non-constrained (left) and constrained (right)

Figure 2.13 illustrates the contribution of the CNEM (right) with respect to the NEM (left) close to the non-convex edge. In Figure 2.5, the diameter of the full red discs represents the value of the shape functions in x.

2.4. Mixed approximations of natural neighbor type

From the above discussion, we conclude that it is possible to derive an interpolation of a scalar field (or components of any tensorial field), due to the approximation of natural neighboring given in equation [2.6]. However, it is well known that in weak formulations that use several unknown fields (mixed formulations), the use of this same interpolation for all the fields leads to instabilities. To avoid this instability in finite elements, interpolations of various orders for each unknown field were proposed in order to verify the stability condition known as the LBB condition

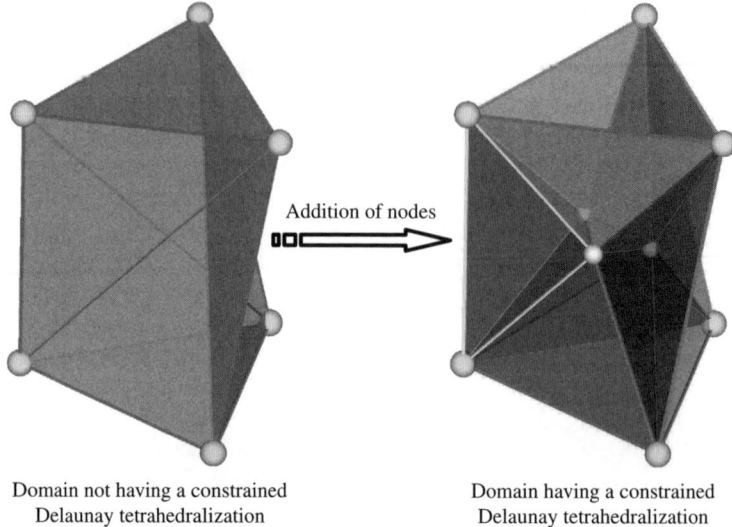

Domain not having a constrained   Domain having a constrained
Delaunay tetrahedralization   Delaunay tetrahedralization

Figure 2.12. *Constrained Delaunay tetrahedralization of the Schönhardt polyhedron*

(proposed originally for the mixed formulations associated with incompressible models). We can immediately understand where the problem is when we proceed with the natural elements as there is only one form of approximation given by equation [2.6].

In this section, we revisit this problem and propose various alternatives to solve it.

2.4.1. *Considering the restriction of incompressibility*

Here, we focus on problems related to incompressible fluids or solids in order to try to define stable mixed approximations. Although several researchers have confirmed that meshless methods are not sensitive to the problem of locking due to incompressibility constraints, this assertion is far from being true.

Thus, we consider a mechanical model defined below:

– Equilibrium equation (the volumic forces and inertia are neglected):

$$\nabla \cdot \boldsymbol{\sigma} = \mathbf{0} \quad \text{in } \Omega. \tag{2.20}$$

– Incompressibility:

$$\nabla \cdot \mathbf{u} = 0 \quad \text{in } \Omega, \tag{2.21}$$

where $\boldsymbol{\sigma}$ is the Cauchy stress tensor and $\mathbf{u}$ is the vector of displacements (or velocities).

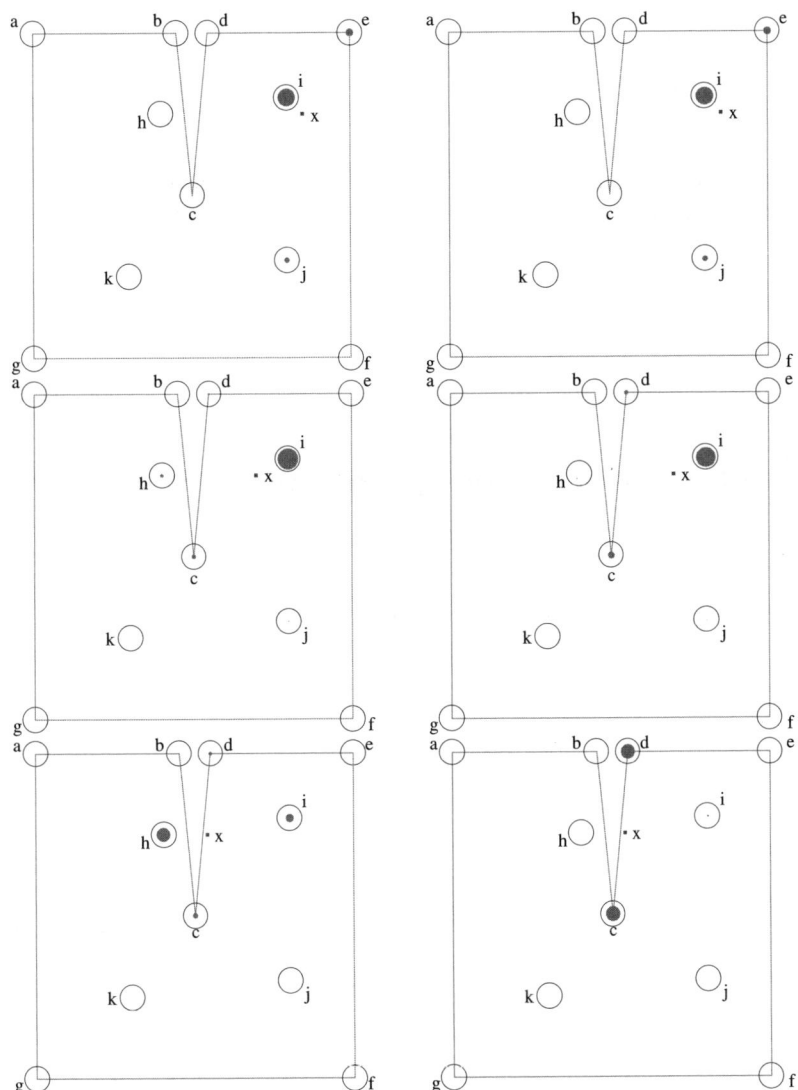

Figure 2.13. *Sibson shape function close to the non-convex edge: NEM (left) CNEM (right)*

The boundary conditions are given by

$$\boldsymbol{\sigma} \cdot \mathbf{n} = \bar{\boldsymbol{t}} \quad \text{on } \Gamma_t, \qquad\qquad [2.22]$$

$$\mathbf{u} = \bar{\mathbf{u}} \quad \text{on } \Gamma_u, \qquad\qquad [2.23]$$

where $\mathbf{n}$ represents the normal unit vector at the boundary $\Gamma = \Gamma_u \cup \Gamma_t$, with $\Gamma_u \cap \Gamma_t = \emptyset$.

If we consider the linear and isotropic case (elasticity under the assumption of small displacements and strains), the constitutive law is written as

$$\boldsymbol{\sigma} = -p\boldsymbol{I} + 2\mu\nabla_s\mathbf{u}, \qquad\qquad [2.24]$$

$$0 = \nabla \cdot \mathbf{u} - \frac{p}{\lambda}, \qquad\qquad [2.25]$$

where p is the hydrostatic component of the stresses and ∇_s is the symmetric component of the gradient.

The coefficients of Lamé λ and μ can be written starting from Young's modulus E and Poisson's ratio ν:

$$\mu = \frac{E}{2(1+\nu)}, \quad \lambda = \frac{2\mu\nu}{1-2\nu}. \qquad\qquad [2.26]$$

When ν moves toward 0.5, λ moves toward infinity, and thus equation [2.25] expresses the restriction of incompressibility $\nabla \cdot \mathbf{u} = 0$.

The variational form of this problem is written:

Look for $\mathbf{u} \in \mathcal{U}$ *such that*

$$\int_{\Omega(t)} \boldsymbol{\sigma} : \varepsilon^* d\Omega = \int_{\Gamma_t} \bar{\boldsymbol{t}} \cdot \mathbf{u}^* d\Gamma \quad \forall \mathbf{u}^* \in \mathcal{V}, \qquad\qquad [2.27]$$

$$\int_{\Omega(t)} \left(-\nabla\mathbf{u} + \frac{1}{\lambda}p \right) p^* d\Omega = 0 \quad \forall p^* \in L_2(\Omega(t)), \qquad\qquad [2.28]$$

where $\mathcal{U} = \{\mathbf{u}|\mathbf{u} \in H^1(\Omega), \mathbf{u}|_{\Gamma_u} = \bar{\mathbf{u}}\}$, $\mathcal{V} = \{\mathbf{u}^*|\mathbf{u}^* \in H^1(\Omega), \mathbf{u}|_{\Gamma_u} = 0\}$. H^1 and L_2 are the usual spaces of Sobolev and Lebesgue, respectively.

If we approximate the displacements and the pressures using a base of finite size:

$$\mathbf{u}^h(\boldsymbol{x}) = \sum_{I=1}^{n} \phi_I(\boldsymbol{x})\mathbf{u}_I, \qquad\qquad [2.29]$$

$$p^h(\boldsymbol{x}) = \sum_{I=1}^{n} \psi_I(\boldsymbol{x})p_I. \qquad\qquad [2.30]$$

This leads to the following linear system after integration

$$
\begin{pmatrix} \overline{K} & G \\ G^T & M \end{pmatrix} \begin{pmatrix} u \\ p \end{pmatrix} = \begin{pmatrix} f \\ 0 \end{pmatrix}.
\tag{2.31}
$$

In the equations defining the functional approximations, n represents the number of nodes considered in the approximation, and $\psi_I(x)$ and $\phi_I(x)$ are the shape functions which we assume to be calculated starting from a certain interpolation of the natural neighbor type.

In equation [2.31],

$$
\overline{K}_{IJ} = \int_\Omega B_I^T \overline{C} B_J \mathrm{d}\Omega,
\tag{2.32}
$$

$$
G_{IJ} = - \int_\Omega \tilde{B}_I^T \psi_J \mathrm{d}\Omega,
\tag{2.33}
$$

$$
M_{IJ} = -\frac{1}{\lambda} \int_\Omega \psi_I \psi_J \mathrm{d}\Omega,
\tag{2.34}
$$

$$
f_I = \int_{\Gamma_t} \phi_I \bar{t} \mathrm{d}\Gamma,
\tag{2.35}
$$

and

$$
\tilde{B}_I = \begin{bmatrix} \phi_{I,1}(x) & \phi_{I,2}(x) \end{bmatrix},
\tag{2.36}
$$

$$
B_I = \begin{pmatrix} \phi_{I,1}(x) & 0 \\ 0 & \phi_{I,2}(x) \\ \phi_{I,2}(x) & \phi_{I,1}(x) \end{pmatrix},
\tag{2.37}
$$

$$
\overline{C}_{IJKL} = \mu(\delta_{IK}\delta_{JL} + \delta_{IL}\delta_{JK}).
\tag{2.38}
$$

In the completely incompressible case $M = 0$. Techniques based on penalization are obtained if we employ a sufficiently large λ parameter.

It is well known that certain approximations do not lead to stable and convergent methods (see, for example, [BAT 96]). The requirement to make these approximations convergent and stable is known under the name of *inf-sup* or LBB condition [BAB 73, BRE 74]. The LBB condition can be written as

$$
\inf_{p^h \in \mathcal{P}_h} \sup_{u^h \in \mathcal{U}_h} \frac{\int_\Omega p^h \nabla \cdot u^h \mathrm{d}\Omega}{||p^h||_0 ||u^h||_1} = \gamma_h \geq \gamma > 0,
\tag{2.39}
$$

where γ is a positive constant independent of the mesh parameter h and $\mathcal{U}_h$ and $\mathcal{P}_h$ being spaces of approximation of displacements and pressures, respectively,

$$|| \cdot ||_0^2 = \int_\Omega (\cdot)^2 d\Omega, \tag{2.40}$$

$$|| \cdot ||_1^2 = \int_\Omega \sum_{i,j=1}^{2} \left(\frac{\partial (\cdot)_i}{\partial x_j} \right)^2 d\Omega. \tag{2.41}$$

It is very difficult to verify this condition analytically. Thus, normal practice proceeds by verifying it numerically, as we will do in section 2.4.2.

2.4.2. Mixed approximations in the Galerkin method

We develop mixed, stable approximations of the natural neighbor type for their use in the discretization of equation [2.27]. The simplest solution is to use an approximation of displacements of a higher degree of consistency than that used for the approximation of pressures (as in the case of the finite elements).

In [SUK 98a, CUE 01], a mixed approximation based on a displacement interpolation by Sibson and pressures by Thiessen (piece-wise constant) was proposed.

$$\mathbf{u}^h(\mathbf{x}) = \sum_{I=1}^{n} \phi_I(\mathbf{x})\mathbf{u}_I, \tag{2.42}$$

$$p^h(\mathbf{x}) = \sum_{I=1}^{n} \psi_I(\mathbf{x})p_I, \tag{2.43}$$

where $\mathbf{u}_I$ and p_I represent displacements and nodal pressures. This approximation was also analyzed in [CUE 03]. Although this mixed approximation gave very good performances in the simulation of the Newtonian and non-Newtonian complex flows [MAR 03], we prove later that it does not verify the LBB condition.

To test the LBB condition numerically, [BAT 96, CHA 93] proposed a numerical test to be fulfilled on several meshes. An equivalent, discrete, expression of equation [2.39] is given as follows:

$$\inf_{\boldsymbol{w}_h \in \mathcal{P}_h} \sup_{\boldsymbol{u}_h \in \mathcal{U}_h} \frac{\mathbf{W}_h^T \boldsymbol{G}_h \mathbf{U}_h}{\sqrt{\mathbf{W}_h^T \boldsymbol{G}_h \mathbf{W}_h} \cdot \sqrt{\mathbf{U}_h^T \boldsymbol{S}_h \mathbf{U}_h}} = \gamma_h \geq \gamma > 0, \tag{2.44}$$

where $\mathbf{W}_h$ and $\mathbf{U}_h$ are the vectors containing the nodal displacements associated with w_h and u_h. $\boldsymbol{G}_h$ and $\boldsymbol{S}_h$ are the matrices associated with the norms:

$$||p_h||_0^2 = \mathbf{W}_h^T \boldsymbol{G}_h \mathbf{W}_h, \qquad [2.45]$$

$$||\mathbf{u}_h||_1^2 = \mathbf{U}_h^T \boldsymbol{S}_h \mathbf{U}_h, \qquad [2.46]$$

$\boldsymbol{G}_h$ and $\boldsymbol{S}_h$ are semi-definite positive and definite positive, respectively [CHA 93]. Thus, it is possible to prove that the first non-zero eigenvalue λ_k of the problem

$$\boldsymbol{G}_h \phi_h = \lambda \boldsymbol{S}_h \phi_h \qquad [2.47]$$

is connected to the sought value γ_h by expression:

$$\gamma_h = \sqrt{\lambda_k}. \qquad [2.48]$$

If the number of degrees of freedom in pressure is n_p and in displacements is n_u, the number of non-physical modes in pressure is given by

$$k_{pm} = k - (n_u - n_p - 1). \qquad [2.49]$$

Bathe [BAT 96] proposed using two or three meshes to test the approximation. If γ_h is not limited by 0, the condition is violated. Moreover, this test is very simple to implement. Thus, we prove with this simple test that the mixed approximation of Sibson–Thiessen does not verify the LBB condition. However, no problem of locking has ever been detected [SUK 98a, MAR 03, CUE 03, CUE 01].

From these facts, we can expect that an enrichment of the approximation of displacements leads to the fulfillment of the LBB condition. However, the Sibson approximations have only linear consistency [SUK 98b]. An approximation of the natural neighbor type exists, which is able to reproduce second-order polynomials [SUK 98a], but it uses the derivative of the unknown function and requires a large CPU cost. For these reasons, we prefer an enrichment based on the paradigm of the partition of unity [BAB 96]. This approach had been already used in [WEL 02].

2.4.3. *Natural neighbor partition of unity*

2.4.3.1. *Partition of unity method*

The main idea of the partition of unity method (PUM) [BAB 96] is to derive approximations enriched by adding functions to the approximation space. Let $\{\Omega_I\}$ be an open cover of the field Ω, $\{\phi_I\}$ be a partition of unity defined on this cover, and $V_I \subset H^1(\Omega_I \cap \Omega)$. Space V is defined by [BAB 96]

$$V := \sum_I \phi_I V_I = \left\{ \sum_I \phi_I v_I \mid v_I \in V_I \right\} \subset H^1(\Omega). \qquad [2.50]$$

Spaces V_I are known to approximate the solution well in any sense.

In the partition of unity method, the main property is that the resulting approximation space inherits the properties from local approximation spaces V_I and the regularity of the partition of unity $\{\phi_I\}$ [BAB 96]. The same idea was proposed independently by Duarte and Oden [DUA 96], where the partition of unity was built by moving least squares and the enrichment consisted of adding polynomials.

With the same approach, and considering that the finite element shape functions define a partition of unity, [ODE 98, STR 01] proposed an enrichment by adding other polynomial terms to the shape functions.

In the following, we propose to enrich the interpolants by Sibson (or Laplace) in order to reproduce polynomials of a higher order, for the approximation of displacements within the framework of the mixed formulations.

We add some new functions to the approximation given by

$$N_{I\alpha} := \phi_I L_\alpha, \qquad [2.51]$$

where L_α is a polynomial of degree p. Because the interpolants of the natural neighbor type have linear consistency, we obtain

$$\sum_I N_{I\alpha} = \sum_I \phi_I L_\alpha = L_\alpha \sum_I \phi_I = L_\alpha, \qquad [2.52]$$

which proves that the L_α polynomials are represented by the approximation constructed in this manner.

On the other hand, given the linear consistency of the interpolant of Sibson (or Laplace) constants $a_{xI}, a_{yI}, I = 1, \ldots, n$ exist that verify

$$\sum_I a_{xI} \phi_I = x, \qquad [2.53]$$

$$\sum_I a_{yI} \phi_I = y. \qquad [2.54]$$

In this manner, if we take $L_\alpha = x$ or $L_\alpha = y$, we have linear dependencies in the approximation basis. When these dependencies cannot be avoided, a suitable technique has to be used to solve the resulting system of equations [DUA 00].

If we note the system of linear equations [2.31], which stems from the application of the Galerkin method by $\tilde{A}\tilde{d} = \tilde{p}$, where $\tilde{A}$ is semi-definite positive, we can apply an iterative algorithm that perturbs the coefficient matrix in order to eliminate its possible singularity (because of the existence of approximation functions which are not linearly independent).

Thus,

$$A = T\tilde{A}T, \tag{2.55}$$

$$d = T^{-1}\tilde{d}, \tag{2.56}$$

$$p = T\tilde{p}, \tag{2.57}$$

with

$$T_{IJ} = \frac{\delta_{IJ}}{\sqrt{\tilde{A}_{IJ}}}. \tag{2.58}$$

The diagonal components of the coefficient matrix of the resulting system $Ad = p$ are all unitary. This matrix is now perturbed:

$$A_\varepsilon = A + \varepsilon I \quad \varepsilon > 0, \tag{2.59}$$

and the resulting matrix A_ε is no longer singular. To solve the initial system, the following iterative procedure is put into place:

$$d_0 = A_\varepsilon^{-1}p, \tag{2.60}$$

$$r_0 = p - Ad_0. \tag{2.61}$$

If $e_0 = d - d_0$, then

$$A_\varepsilon e_0 \simeq Ae_0 = Ad - Ad_0 = r_0 \tag{2.62}$$

The algorithm consists of successively calculating

$$r_i = r_0 - \sum_{j=0}^{i-1} Ae_j, \tag{2.63}$$

$$e_i = A_\varepsilon^{-1}r_i, \tag{2.64}$$

$$d_i = d_0 + \sum_{j=0}^{i-1} e_j. \tag{2.65}$$

until convergence. The final solution is given by $\tilde{d} = Td$. In practice, a perturbation $\varepsilon = 10^{-10}$ is sufficient.

2.4.3.2. Enrichment of the natural neighbor interpolants

Here, we consider various enrichments of spaces $u - p$, which will be tested to conclude their relevance.

The objective is to keep the distance limited between the exact and approached solutions:

$$\|u - u^h\| \le c \cdot d(u, \mathcal{U}_h), \tag{2.66}$$

where c is a constant that is independent of the meshing parameter h (average distance between the nodes in the meshless techniques) and of the value λ (which move toward infinity when we approach the incompressibility).

When it becomes increasingly large, consequently the quantity $\|\text{div}\boldsymbol{u}^h\|$ must decrease to maintain the limited pressure. In this manner, the space of approximated displacements $K(0) = \{\boldsymbol{u}^h | \boldsymbol{u}^h \in \mathcal{U}_h, \ \ \text{div}\boldsymbol{u}^h = 0\}$ should be as rich as possible. For this, we test some enrichments of the approximation of displacements while preserving that of Sibon or Thiessen for the pressures.

2.4.3.2.1. Sibson–Thiessen mixed formulation

The simplest of the mixed formulations is constructed, as we have already discussed in section 2.4.2, with the Sibson interpolation for displacements and the Thiessen interpolation for pressures. The results of the numerical test to estimate the verification of the LBB condition were published in [GON 04a]. The approximation space of displacements is not rich enough to verify the LBB condition.

2.4.3.2.2. Approximation of displacements enriched with polynomials $\{1, x^2, y^2\}$

We add the polynomials $\{1, x^2, y^2\}$ to the approximation base of displacements, which thus becomes:

$$\Phi = \phi_I \times \{1, x^2, y^2\}, \quad I = 1, \ldots, n, \qquad [2.67]$$

where n is the number of nodes. It is simple to check the relation:

$$\text{span}\{\Phi\} = \text{span}\{1, x, y, x^2, y^2, x^3, y^3, x^2y, xy^2\}. \qquad [2.68]$$

This particular enrichment leads to an incomplete polynomial space of order 3. It was selected to avoid the dependencies between the base functions. On the other hand, the resulting approximation contains parasitic modes in pressure. It is recommended at least to completely reproduce the bilinear base. This conclusion is similar to that given in [WEL 02].

2.4.3.2.3. Approximation of displacements enriched with polynomials $\{1, x, y, xy\}$

Now, the enrichment of the approximation of displacements is made by starting from the monomials $\{1, x, y, xy\}$.

The resulting approximation space is given by

$$\Phi = \phi_I \times \{1, x, y, xy\}, \quad I = 1, \ldots, n. \qquad [2.69]$$

Hence,

$$\text{span}\{\Phi\} = \text{span}\{1, x, y, xy, x^2, y^2, x^2y, xy^2\}. \qquad [2.70]$$

However, as shown in the following, we find the problem of the dependence between the base functions.

Because the Sibson or Laplace interpolation can reproduce the linear functions, such constants $a_I^x, a_I^y, I = 1, \ldots, n$ exist, such as $\forall \boldsymbol{x} \in \Omega$,

$$\sum_{I=1}^{n} \phi_I(\boldsymbol{x}) = 1,$$ [2.71]

$$\sum_{I=1}^{n} a_I^x \phi_I(\boldsymbol{x}) = x,$$ [2.72]

$$\sum_{I=1}^{n} a_I^y \phi_I(\boldsymbol{x}) = y,$$ [2.73]

where n is the number of natural neighbors at the point considered.

In this manner, $\{1, x, y\} \subset \mathrm{span}\{\Phi\}$. If the partition of unity is used now,

$$\sum_{I=1}^{n} (\phi_I x) = x \sum_{I=1}^{n} \phi_I = x.$$ [2.74]

With equation [2.72], we have:

$$\sum_{I=1}^{n} (\phi_I x) - \sum_{I=1}^{n} a_I^x \phi_I = 0,$$ [2.75]

which are then linearly dependent. The same reasoning can be applied finally in y. Finally,

$$\sum_{I=1}^{n} a_I^x (\phi_I y) = y \sum_{I=1}^{n} a_I^x \phi_I = xy,$$ [2.76]

and consequently,

$$\sum_{I=1}^{n} a_I^x (\phi_I y) - \sum_{I=1}^{n} a_I^y (\phi_I x) = 0.$$ [2.77]

In spite of these dependencies, the algorithm presented in section 2.4.3.1 can be applied in order to solve the resulting system. Various shape functions, associated with the central node in Figure 2.14, are represented in Figure 2.15(a), (b), and (c). Contrary to the preceding approximations, this approximation verifies the LBB condition.

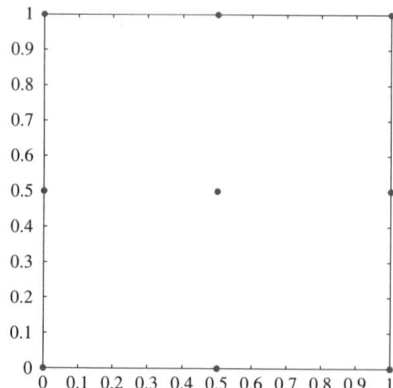

Figure 2.14. *Cloud of nine regularly distributed nodes*

2.4.3.2.4. Approximation of displacements enriched with polynomials $\{1, x, y, xy\}$ combined with a Sibson interpolation of pressures

This formulation is similar to the previous approximations except the pressure is now built up from the Sibson interpolant.

2.4.3.2.5. Approximation of displacements enriched with the monomials $\{1, xy\}$

The enrichment of the approximation of displacements with $\{1, xy\}$ avoids the existence of dependencies between the resulting approximation functions while verifying the LBB condition.

$$\Phi = \phi_I \times \{1, xy\}, \quad I = 1, \ldots, n. \tag{2.78}$$

Hence,

$$\text{span}\{\Phi\} = \text{span}\{1, x, y, xy, x^2y, xy^2\}. \tag{2.79}$$

This formulation greatly resembles the approximation of the MINI finite element [ARN 84] (linear triangular element enriched with a bubble for displacements and linear for pressures). This formulation verifies the LBB condition for an interpolation of the pressures of Sibson or Thiessen [GON 04a].

Among all the possibilities that we will discuss, the latter seems to be the most suitable because it enables us to check the LBB condition without inducing singularities in the system of equations to be solved. Moreover, the fewer the enrichment functions, the fewer additional unknowns are introduced into the discrete model.

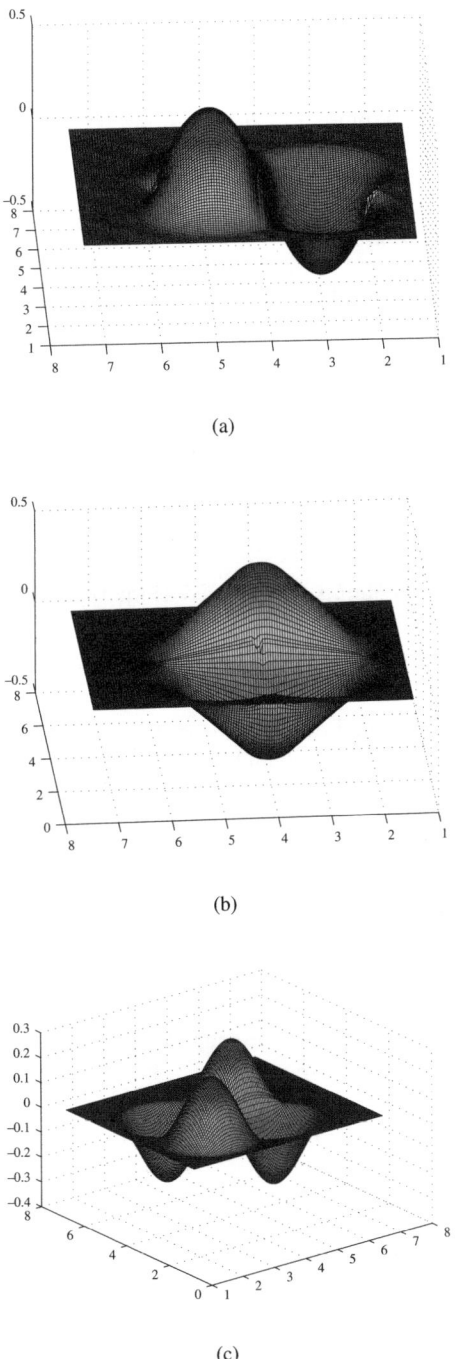

(a)

(b)

(c)

Figure 2.15. *Shape functions ϕx (a), ϕy (b), and ϕxy (c)*

2.5. High order natural neighbor interpolants

We presented earlier an interpolation based on the natural neighbors (equation [2.6]), which is consistent at order 1 (linear consistency) and continuous (C^0). In this section, we explore a path that allows us to generate natural neighbor interpolants of arbitrary consistence and continuity.

2.5.1. Hiyoshi–Sugihara interpolant

Recently, Hiyoshi and Sugihara proposed a generalization of natural neighbor coordinates of a plot cloud [HIY 99]. Before discussing this generalization, we define the concept of the Laguerre–Voronoi diagram (weighted Voronoi diagram).

Let $B = \{b_1, b_2, \ldots, b_n\}$ be a set of closed spheres of $\mathbb{R}^d$, that is, $b_I = (\boldsymbol{x}_I, \varrho_I)$, where $\boldsymbol{x}_I$ is the center of the sphere and ϱ_I represents its radius. The *weighted distance* from a point to the sphere is defined as

$$\pi_I(\boldsymbol{x}) = ||\boldsymbol{x} - \boldsymbol{x}_I|| - \varrho_I. \qquad [2.80]$$

The Laguerre–Voronoi diagram is thus defined as the space decomposition into areas of the shape:

$$V_I = \{\boldsymbol{x} \in \mathbb{R}^d : \pi_I(\boldsymbol{x}) < \pi_J(\boldsymbol{x}) \, \forall J \neq I\}. \qquad [2.81]$$

Thus, the Laguerre–Voronoi diagram shares the major properties of the standard Voronoi diagrams, but not all. For example:

– the Laguerre–Voronoi cell associated with a node does not exist necessarily,

– if it exists, it is convex,

– the Laguerre–Voronoi facet associated with two neighboring nodes is perpendicular to the segment connecting these two nodes.

The proof of these properties is in [HIY 99].

The key point in the definition of the generalized coordinates is the consideration of a set of nodes fitted with the spheres which have increasing radii. It should be stressed that the Laguerre–Voronoi diagram for $\varrho_I = 0$, $\forall I$, becomes that of Voronoi. If, starting from 0, all the radii of the spheres are increased, except for that associated with the point of evaluation, the Laguerre–Voronoi cell at the point of evaluation will necessarily be reduced until disappearance for a sufficiently large value of the radii.

In Figure 2.16, a Laguerre–Voronoi cell associated with the point of evaluation $\boldsymbol{x}$ (surrounded by four nodes $\boldsymbol{x}_1$ to $\boldsymbol{x}_n$) is represented. In this figure, the dotted line represents the reduced cell associated with the original cell (the latter is represented by a continuous line) when the weight $\varrho \neq 0$.

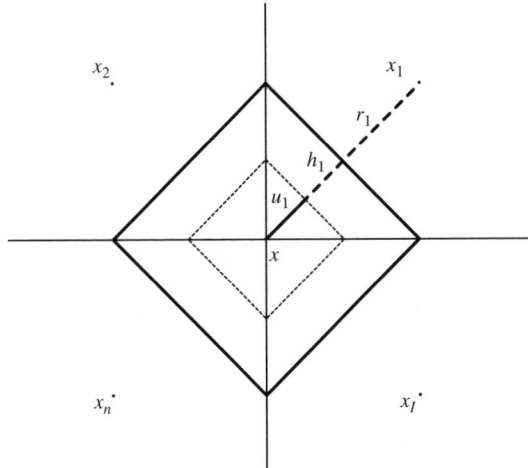

Figure 2.16. *Cells being reduced in the Laguerre–Voronoi diagram*

The distance between the point of evaluation and node x_I is represented as r_I, whereas the distance between this point and the $(n-1)$-dimensional Voronoi facet is represented as h_I, as in Figure 2.4. This same distance will be represented as u_I in the Laguerre–Voronoi diagram.

According to these definitions, we can introduce the following coordinates, on a cell being reduced and associated with a point of evaluation x:

$$\zeta_I^k(x) = r_I \int_0^{u_J} \zeta_I^{k-1}(x)du_J', \qquad [2.82]$$

with

$$\zeta_I^0(x) = \frac{s_I}{u_I}, \qquad [2.83]$$

s_I being the $(d-1)$-dimensional measure of the Laguerre–Voronoi facet and x_J as one of the natural neighbors of node x_I. du_J' represents a dummy variable for the integration on the segments u_I, as in Figure 2.16. It is possible to normalize this coordinate to obtain

$$\psi_I^k(x) = \frac{\zeta_I^k(x)}{\sum_{J=1}^n \zeta_J^k(x)}, \qquad [2.84]$$

where the superscript k represents the number of times that the shape function obtained can be derived [HIY 99]. In the earlier definition, the functions $\psi_I^0(x)$ are only defined when x does not coincide with any node x_I. In such a case,

$$\psi_I^0(x_J) = \delta_{IJ}. \qquad [2.85]$$

these coordinates generalize the Laplace–Sibson sequence and allow us to interpolate the data with smoother shape functions. Some interesting properties of these generalized coordinates are as follows [HIY 99]:

 – the Kronecker delta property:

$$\psi_I^k(\pmb{x}_J) = \delta_{IJ};$$

 – continuity $\mathcal{C}^k$ if $\pmb{x}$ belongs to the Delaunay spheres of the nodes. Continuity $\mathcal{C}^0$ at the nodes was not proven analytically, but the numerical tests carried out by Hiyoshi and colleagues [HIY 99] seem to prove this continuity. Finally, continuity $\mathcal{C}^\infty$ is obtained at any other point.

In Figure 2.17, the shape functions with continuity $\mathcal{C}^1$ and continuity $\mathcal{C}^2$ are represented for a set of regularly distributed nodes. This set of interpolation functions

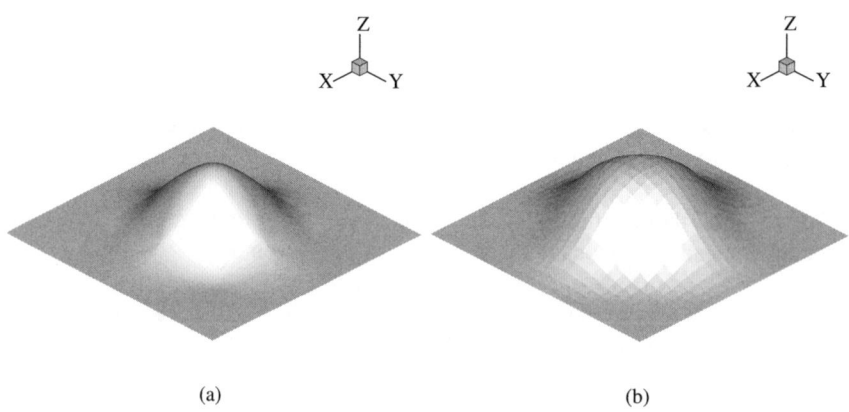

(a)   (b)

Figure 2.17. *Hiyoshi–Sugihara interpolant with continuity* $\mathcal{C}^1$ *(a) and continuity* $\mathcal{C}^2$ *(b)*

have only linear consistency. This can be easily proven if we consider the limiting situation of only three nodes next to a point of evaluation: it is impossible to define a quadratic surface by prescribing only three points in the space.

Thus, the generalization of the natural neighbor coordinates must also examine the question of increasing consistency. We describe a recent technique based on the use of the De Boor algorithm [FAR 02]. This algorithm is the same as that used in the B-spline curves to build curves of a higher order by a linear combination of linear interpolants.

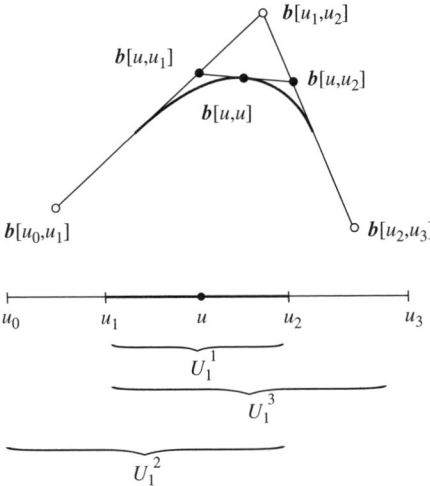

Figure 2.18. *De Boor algorithm*

2.5.2. *The De Boor algorithm for B-splines*

The "de Casteljau" algorithm for Bezier curves establishes that such curves can be obtained by successive applications of linear interpolations [FAR 02], that is, if points $b_0, b_1, \ldots, b_n \in \mathbb{E}^3$ and $t \in \mathbb{R}$ are given, the construction

$$b_i^r(t) = (1-t)b_i^{r-1}(t) + tb_{i+1}^{r-1}(t) \text{ with } \begin{cases} r = 1, \ldots, n \\ i = 0, \ldots, n-r \end{cases} \quad , \qquad [2.86]$$

where $b_i^0 = b_i$, defines the expected Bezier curve.

The de Boor algorithm generalizes the algorithm by introducing a parametric space represented by an arbitrary sequence of nodes u_0, u_1, u_2, u_3. A quadratic Bezier curve can thus be parameterized, for example, by the sequences $0, 0, 1, 1$. The quadratic tranform $b[u, u]$ can then be written as (see Figure 2.18)

$$b[u, u] = \frac{u_2 - u}{u_2 - u_1} b[u_1, u] + \frac{u - u_1}{u_2 - u_1} b[u, u_2]$$

$$= \frac{u_2 - u}{u_2 - u_1} \left(\frac{u_2 - u}{u_2 - u_0} b[u_0, u_1] + \frac{u - u_0}{u_2 - u_0} b[u_1, u_2] \right) \qquad [2.87]$$

$$+ \frac{u - u_1}{u_2 - u_1} \left(\frac{u_3 - u}{u_3 - u_1} b[u_1, u_2] + \frac{u - u_1}{u_3 - u_1} b[u_2, u_3] \right).$$

The key point of the de Boor algorithm is that it expresses u in terms of intervals of increasing size. The B-spline curves result from the union of segments of polynomial curves. The key aspect of the de Boor algorithm is that it expresses u in terms of intervals of growing size. B-spline curves consist of a union of polynomial curve segments. Following the notation in [FAR 02], let U be an interval $[u_I, u_{I+1}]$ in the sequence of knots. Then, there will be an ordered sequence of knots U_i^r, each containing u_I or u_{I+1}, such that U_i^r consists of $r + 1$ successive knots and u_I is the $(r - i)$th element of U_i^r.

A degree n curve segment corresponding to the interval U is then given by $n + 1$ control points d_i. Each intermediate control polygon leg d_i^r, d_{i+1}^r can then be viewed as an affine image of U_{i+1}^{n-r+1}. The point d_i^{r+1} is the image of u under such an affine map.

It is well-known that a non-parametric B-spline function $d(u)$ can be written as a parametric curve with the control points [FAR 02]:

$$d_i = \begin{bmatrix} \xi_i \\ d_i \end{bmatrix}, \quad \text{with } i = 0, \dots, L,$$

and $L = K - n + 1$, with K the number of nodes and n the degree of the curve. In this case, points ξ_i are called the *Greville abscissae* and are given by

$$\xi_i = \frac{1}{n}(u_i + \dots + u_{i+n-1}).$$

For $n = 2$, it is easy to prove that the *Greville abscissae* coincides with the Voronoi vertices of the node sequences.

If we proceed in a non-parametric form and use the equivalence between the Sibson interpolation and the linear interpolation in the unidimensional case, this algorithm can be obtained by application of the Sibson interpolation on U_i^r in which we eliminate $r - 1$ from the closest neighbors of the point u:

$$
\begin{aligned}
b[u, u] &= \phi_1(u)b[u_1, u] + \phi_2(u)b[u, u_2] \\
&= \phi_1(u)\left(\varphi_0^2(u)b[u_0, u_1] + \varphi_2^2(u)b[u_1, u_2]\right) \\
&\quad + \phi_2(u)\left(\varphi_1^2(u)b[u_1, u_2] + \varphi_3^2(u)b[u_2, u_3]\right),
\end{aligned}
\qquad [2.88]
$$

where $\phi_I(u)$ represents the coordinates of natural neighboring of the point u compared to node I but calculated on the interval U_i^r, that is, by eliminating the $r - 1$ natural neighbors from the interval. The notation used is illustrated in Figure 2.19.

2.5.3. *B-spline surfaces and natural neighboring*

2.5.3.1. *Some definitions*

The De Boor algorithm can be extended to several dimensions. In the following, we use the Sibson coordinates, even if the proposed algorithm can also be applied in the case of the Laplace or Hiyoshi–Sugihara interpolants.

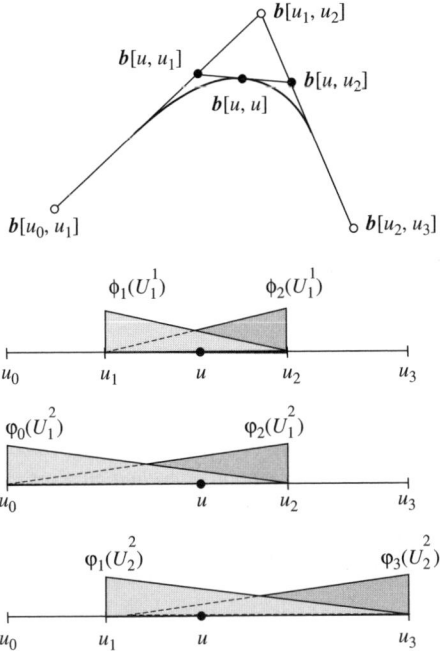

Figure 2.19. *Schematic representation of the de Boor algorithm employing natural neighbors. The domain of each function is given within parentheses*

Let us consider a set of nodes $N = \{x_1, x_2, \ldots, x_M\} \subset \mathbb{R}^2$ and a quadratic surface (the extension to the 3D case is immediate). In the same way, we restrict ourselves to the non-parametric case.

We can define a new surface class by

$$s(x) = \sum_{I=1}^{n} \sum_{J=1}^{n^I} N_{IJ}(x)d_{IJ}, \quad \text{with } d_{IJ} = d_{JI}, \qquad [2.89]$$

where n represents the number of neighbors of point x. Moreover,

$$N_{IJ}(x) = \phi_I(x)\varphi_J^I(x), \qquad [2.90]$$

and d_{IJ} represents the control points, if we use the terminology of B-splines (i.e. degrees of freedom). $\phi_I(x)$ represents the Sibson coordinate of point x with respect to node I to I. Functions $\varphi_J^I(x)$ represent the coordinates of natural neighboring of point x with respect to node J, when we elliminate node I from the original set of nodes (Figure 2.20). Finally, n^I is the number of natural neighbors of point x

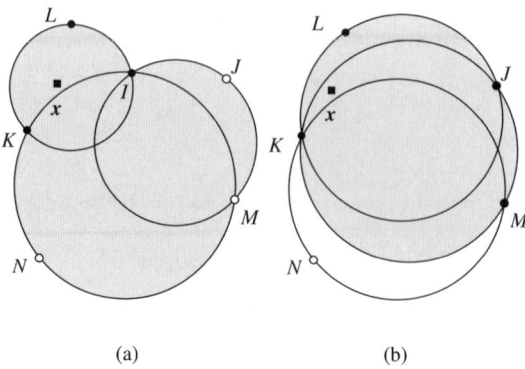

(a)   (b)

Figure 2.20. *Diagrammatic representation of the algorithm proposed: (a) nodes $\{I, \ldots, N\}$. We consider a point of evaluation $\boldsymbol{x}$, of which the natural neighbors are represented by circles – the support of function ϕ_I is indicated. (b) After elimination of node I, the support of function φ_J^I is indicated – the new set of neighboring nodes is composed of $\{J, K, L, M\}$*

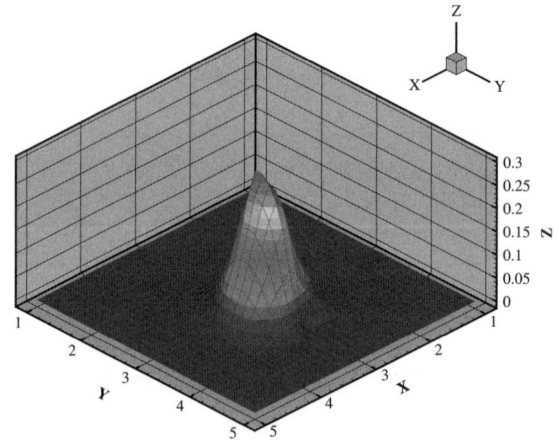

Figure 2.21. *Function N_{IJ} in an irregular distribution of nodes*

when node I is eliminated, as with the De Boor algorithm. It should be stressed that the number of degrees of freedom is much lower than $M^2/2$ because the nodes in equation [2.89] relate only to the natural neighbors of each node.

The typical form of functions N_{IJ} can be evaluated in Figure 2.21 for an irregular distribution of nodes.

2.5.3.2. *Surface properties*

The surfaces defined by equation [2.89] maintain a certain resemblance to the B-spline curves. We quote here some properties of these surfaces by limiting ourselves to the quadratic case for reasons of simplicity (the proof of all these properties can be found in [GON 08]).

PROPERTY 2.5.1. *Functions N_{IJ} are always positive.*

PROPERTY 2.5.2. *Functions N_{IJ} define a partition of unity, that is,*

$$\sum_{I=1}^{n} \sum_{J=1}^{n^I} N_{IJ}(\boldsymbol{x}) = 1 \quad \forall \boldsymbol{x} \in \mathbb{R}^2. \quad\quad [2.91]$$

This property is of great importance when this approximation is used within the framework of Galerkin type discretization because it ensures the reproduction of constant functions (translation of a rigid body). Similarly, we can also prove linear consistency, that is, the reproduction of the linear functions, which, with the partition of unity, ensures the representation of the movements of a rigid body (translation and rotation).

PROPERTY 2.5.3. *Functions $N_{IJ}(\boldsymbol{x})$ can represent any linear polynomial.*

The proof of this property also makes it possible to prove quadratic consistency and, while proceeding by induction, to prove finally the consistency of order p.

PROPERTY 2.5.4. *Functions $N_{IJ}(\boldsymbol{x})$ make it possible to represent any quadratic polynomial.*

To see the proof of consistency of higher order, refer to Farin [FAR 02].

PROPERTY 2.5.5. *The surfaces suggested are C^{p-1}, where p represents the order of the surface except with the nodes and on the segment connecting two neighboring nodes, where they have only one continuity C^0.*

2.5.3.3. *The case of repeated nodes*

It is well known that in the case of B-splines, the continuity of the sequence of curves can be controlled by repeating nodes. For infinitely close nodes (repeated nodes), there appears to be a discontinuity in the derivative, which gives a surface to continuity C^0 which further interpolates the data at the nodes.

PROPERTY 2.5.6. If a node I has a multiplicity r, the resulting surface of order n will have a continuity C^{n-r} everywhere, except on the segments connecting two neighboring nodes where there will be only a continuity C^0.

This last property has a major importance when this type of approximation is used in the context of the Galerkin type discretizations, as is the case within the framework of FEM. The use of repeated nodes ensures the interpolation without considering the consistency of the approximation, allowing a simple imposition of the essential boundary conditions along the boundary of the domain. For this, and for the case within the framework of the FEM, it is sufficient to substitute the known value (on the boundary where the boundary conditions of the Dirichlet type are imposed) in the associated nodal variable.

On the other hand, there is an important property of B-splines which is lost in the construction that we have just proposed. If one (or several) new node(s) is introduced, the resulting surface is modified. This is due to the fact that the Sibson interpolation does not have the property known as *idempotence* [FAR 03].

Chapter 3

Numerical Aspects

The geometrical aspects related to the definition of the natural element method, as well as the non-polynomial character of the associated interpolation functions, require the implementation of specific algorithms. These must be robust and effective. In this chapter, we propose to detail some of the most specific ones.

We begin with the search for natural neighbors of a generic point of the domain (section 3.1), a preliminary search for the calculation of natural element method (NEM) shape functions in this point (section 3.2). We continue with a key point of the passage from the strong forms to the weak forms, which is numerical integration (section 3.3). Indeed, the complexity of the approximation function used, as well as the incompatibility between the domains of integration and the support of the interpolation functions, is at the origin of the inaccuracies in the numerical integration of the variational formulations. We complete by presentation of a NEM approach based on an octree structure (section 3.4) which draws upon interesting prospects for intensive calculation of domains under the assumption of small strains.

3.1. Searching for natural neighbors

The calculation of weak forms associated with various problems that we seek to discuss requires a numerical integration of the form given below.

$$\int_{\Omega} f(\boldsymbol{x}) \mathrm{d}\Omega \approx \sum_{i=1}^{\mathrm{nip}} \omega_i f(\boldsymbol{x}_i). \qquad [3.1]$$

In equation [3.1], ω_i represents weights (which are dependent on the selected method of integration), nip is the number of integration points, and $\boldsymbol{x}_i$ are the points of integration.

It is, however, possible to carry out this integration on the Delaunay triangles in a two-dimensional (2D) case, and on the Delaunay tetrahedrons in a three-dimensional (3D) case. It is a known fact that the use of three points (four for 3D cases) of integration for each triangle (tetrahedron, respectively) leads to integration errors in the NEM. In [GON 04a], a study was carried out on the influence of this error on the results obtained, as well as on the means of remedying it. It was shown that, in spite of the integration error due to the non-polynomial nature of the NEM shape functions, the result obtained is generally of a better accuracy than that obtained with linear finite elements.

Whatever method employed to perform the integration, a search for the neighbors of the integration point must be accomplished. This is equivalent to searching for all the triangles whose circumcircle contains the point x. In that case the three nodes of the triangle will be natural neighbors of the integration point. The naive approach to this problem is to perform an $\mathcal{O}(m^2)$ search (being m the number of nodes in the set, as mentioned in the preceding section), by testing if the evaluation point is a natural neighbor of each node in the set. As will be shown later on, this scheme performs extremely badly, and should be employed only when using very coarse clouds of nodes. In the original work by Braun and Sambridge [BRA 95a], the search for natural neighbors was done by employing the walking triangle algorithm. This method is used to search for the triangle containing the evaluation point.

In our approach this search is not necessary since we employ a method that closely resembles the structure of traditional FE codes. Thus, a "natural element" will be composed by a triangle, which defines, in general, the integration domain, and a number of nodes whose associated shape function's support covers any of the integration points. A similar definition can be given if stabilized conforming nodal integration [CHE 01] is used, for instance, instead of traditional, Gauss-based quadratures.

Recently, a new algorithm has been proposed in [CAI 05] for the natural neighbor search. In it, the possible neighbor candidates are limited to a squared region of length $2r$ around the given integration point. In this way the search is done in an algorithm of order $\mathcal{O}(n \cdot m)$ where $n < m$. Although valid in principle, this algorithm possesses a main drawback derived from the nature of natural neighborhood itself. Two points can be neighbors even if they are far away from each other, depending on the relative position of the other nodes. This circumstance, that constitutes one of the main differences between NEM and other meshless methods, complicates the practical application of the before mentioned algorithm, so that the number of possible neighbor candidates n to be considered should be high enough. Otherwise, the risk of eliminating natural neighbors of the given point is always present, and these will be omitted, thus altering the approximation result.

We have developed a new searching algorithm that performs the search in an "expansive" way, starting with the triangle that defines the integration points. We refer in this explanation to two-dimensional settings for simplicity, although the algorithm is extended to three-dimensional settings straightforwardly. Obviously, the Delaunay triangulation of the points must be previously computed, prior to the application of the algorithm. This is done in our case by employing the "Detri" software [MUE 93], that has proven to be very fast and efficient. It employs a symbolic perturbation technique to avoid degenerate cases [EDE 90]. Nevertheless, this particular choice does not affect the conclusions of the work here presented.

Let us consider, for example, a set of nodes whose triangulation gives 11 triangles as presented in Figure 3.1. The algorithm begins with the triangle defining the domain of integration considered. Here, the triangle is marked as 1. The three nodes of triangle 1 are obviously natural neighbors of the three points of integration of the triangle (stage 1).

Stage 2 (Figure 3.2) then consists of an algorithm moving through the edges of the triangle toward the close triangles, beginning with triangle 2. The circumscribed circle with triangle 2 contains some of the points of integration of element 1; the node of the triangle 2 not belonging to triangle 1 is then added to the list of natural neighbors. This research continues in a recursive way on the basis of triangle 2, leading toward triangle 3. The circumscribed circle with triangle 3 does not contain any point of integration of triangle 1, which makes it possible to stop research in this direction.

With all the triangles surrounding triangle 2 having been traversed, the algorithm then consists of returning to triangle 1, thus moving through its second edge to go toward triangle 4. The method continues in the same manner as for triangle 2, and completes with triangle 7, the last neighbor of triangle 1.

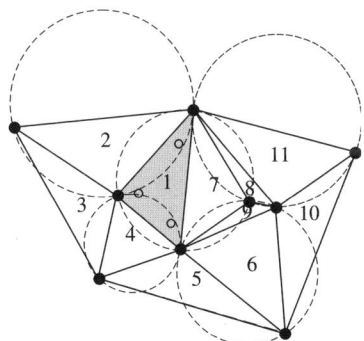

Figure 3.1. *Stage one of the search algorithm – points of integration are represented by small circles*

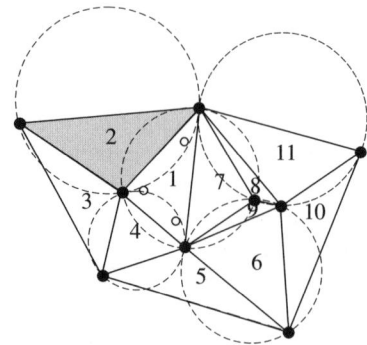

Figure 3.2. *Stage 2 of the search algorithm – displacement toward triangle 2 and search for neighborhood*

Algorithm 1.   Search for natural neighbors for given natural elements:

triangle = initial triangle
for the neighboring triangles (sharing an edge) **do**
   if a point of integration is in the circumscribed circle **and** not "treated" triangle,
     then add again from the new node to the list of the neighbors
     then identify the triangle as "treated"
     then triangle = next neighboring triangle
   if not return to previous triangle
   end if
end for
return

This recursive search algorithm makes it possible to find all the natural neighbors of triangle 1. The algorithm runs in approximate constant time for each triangle, since the number of checked neighboring triangles is nearly constant for each of them. Thus, the total time for the complete neighbor search is $\mathcal{O}(n)$. For this to be true, it is necessary to perform a storage of triangle neighboring a given one. This can be done in the triangulation process, thus saving computational time. We refer the reader to [MUE 93] for further details on the particular data storage algorithm.

The savings in calculation time obtained, thanks to this search by neighboring triangles, are given in Figure 3.3. The saving is of a factor of approximately 4,500 for a mesh comprising approximately 20,000 tetrahedrons.

3.2. Calculation of NEM shape functions of the Sibson type

The calculation of the natural neighbor shape functions (Sibson or Laplace) calls for certain observations as a preamble. In the 3D context, the calculation of NEM shape functions requires the calculation of the volume of Voronoi cells in the case of

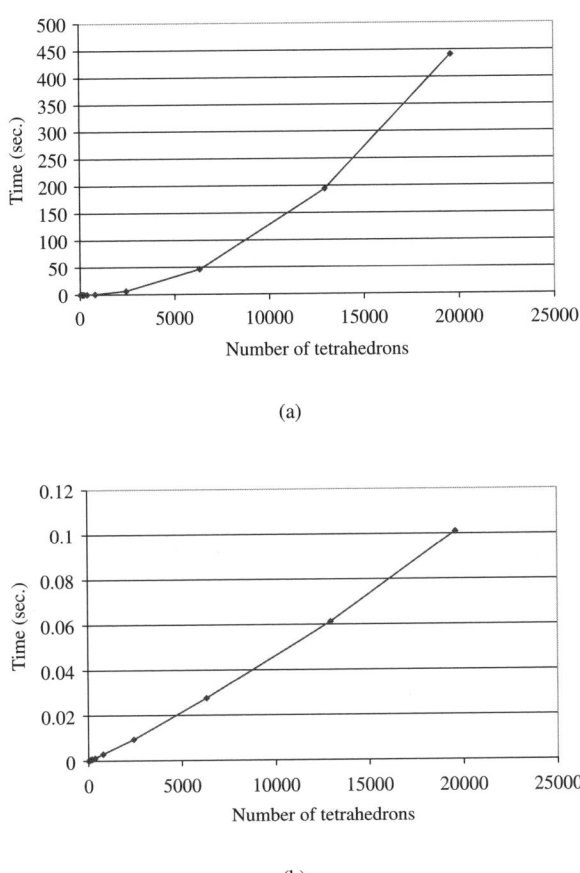

(a)

(b)

Figure 3.3. *Time taken for the search for the neighbors for: (a) a traditional algorithm in* $\mathcal{O}(n^2)$, *(b) the algorithm proposed in* $\mathcal{O}(n)$

the Sibson interpolant, and the areas of the corresponding Voronoi facets (polygons) for Laplace interpolants.

As can be seen in this section, these two calculations have many points in common and can, for certain aspects, be reduced to the same algorithms written in different dimensions.

In practice, the calculation of the Sibson-type NEM (or constrained NEM (CNEM)) shape functions in a point x comprises two main stages. The first stage relates to the local amendment of the Voronoi diagram (possibly constrained),

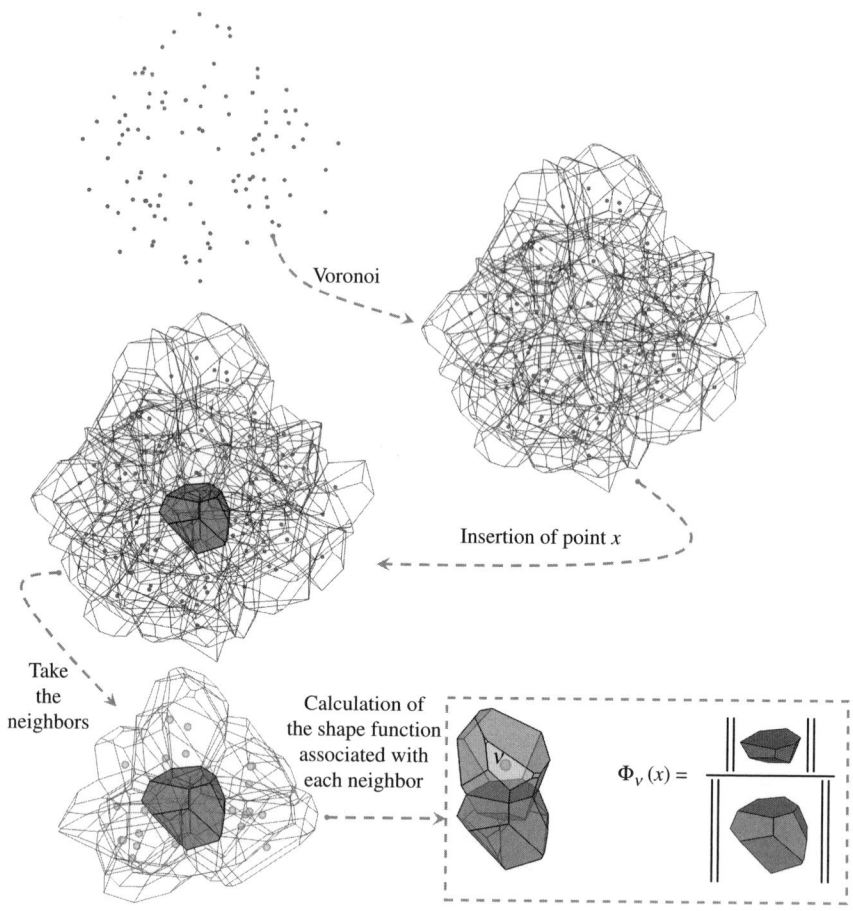

Figure 3.4. *Calculation of Sibson NEM shape functions*

following the addition of point x. This stage is presented in section 3.2.1. The second stage relates to the calculation of the measurement of volume common to the cell associated with point x, and each cell associated with a neighboring node x (in the Voronoi sense in a new diagram) before the insertion of point x. This calculation can be carried out by various approaches. These are presented in section 3.2.2, and their performances are compared in section 3.2.3. An illustration of the complete process is given in Figure 3.4.

A detailed study of the construction and specificity of CNEM shape functions in 3D, in particular in the neighboring non-convex areas of the boundary, is given in [ILL 08].

3.2.1. *Stage-1: insertion of point x in the existing constrained Voronoi diagram (CVD)*

Considering that want to insert a point in an existing diagram, the use of an incremental algorithm is essential. Two incremental algorithms, similar in their approach, are at our disposal [BER 94]: that of Bowyer [BOW 81] which works on the vertices, and that of Watson [WAT 81] which works on the tetrahedrons. The latter was developed for the construction of a *non-constrained* Voronoi diagram, but by using them in an incremental way on an existing (CVD), they maintain the constrained character of the diagram.

Each increment of these two algorithms is broken down into two main stages:

– Look for all the tetrahedrons whose circumscribed sphere contains point x, and break them down. This stage is itself broken down into two other stages:

  - look for tetrahedron which contains point x (see detail just below) and
  - look for all the tetrahedrons whose circumscribed sphere contains point x: this search is local, it is done by "visiting" the tetrahedrons which are near the first.

– Mesh the cavity thus obtained compared to point x, and upgrade the neighboring relationships between the tetrahedrons.

3.2.1.1. *Look for a tetrahedron which contains point x*

For this, it is possible to use the descent in gradient based on the Voronoi graph suggested by Schmitt and Borouchaki [SCH 90], or that based on the graph of Delaunay proposed by Green and Sibson [GRE 78]. In the case of non-convex fields, these two searches can fail if one starts the search starting from a non-visible tetrahedron from point x. Thus, it is necessary to set up other types of searches, such as, for example, "oil spot" or octree searches.

3.2.1.2. *Note concerning the problem of flat tetrahedrons*

Flat tetrahedrons can appear at the time of the meshing of the cavity. This occurs when point x is on the circumscribed circle on a Delaunay tetrahedron face (or if it is very close). According to the tolerance that can be taken for the empty sphere test, we can eliminate the two adjacent tetrahedrons Ta and Tb (Figure 3.5), or only one of them (whereas both must be eliminated). In which case, the face common to the latter will be a face of the cavity, which will consequently create a flat (or quasi flat) Delaunay tetrahedron at the time of meshing, compared to point x. The search for the coordinates of the node associated with such a tetrahedron leads to an indetermination (an infinity of spheres pass by this circle). It is possible to avoid creating such tetrahedrons. For this, it is sufficient to check before remeshing the cavity, if there is one (or several) face(s) of the latter likely to create a flat tetrahedron (point very close to the plane idem formed by the three nodes of the face). If such faces are found, we can only be in the presence of the case previously cited (point x close to the circle circumscribed with the face; the proof is given below), and the adjacent tetrahedron with this face (that which was not removed) can thus be removed,

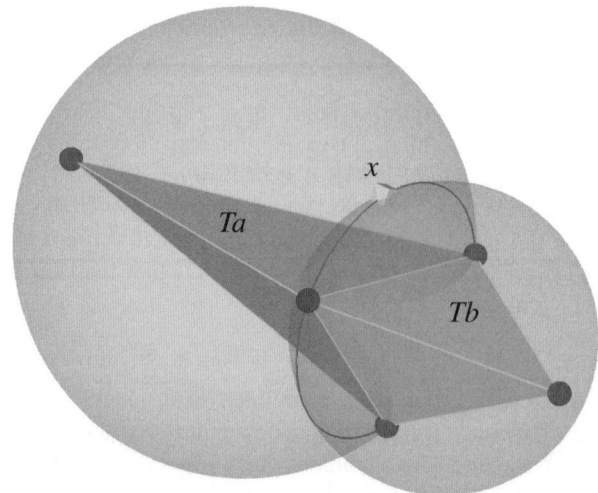

Figure 3.5. *Case of figure enabling the creation of a flat tetrahedron*

something that can be done (without forgetting to upgrade the cavity). It is only after having checked all the faces of the cavity that the meshing can be carried out.

Proof

Let us imagine that there is a face of the cavity for which item x belongs to the plane idem passing by the three nodes of the faces (or if it is very close to it); three cases are possible:

– point x is outside the circumscribed circle at the face,

– point x is inside the circumscribed circle at the face, and

– point x is on the circumscribed circle at the face (or very near to this circle).

The first proposal is impossible because the second tetrahedron sharing the face could not have been eliminated, considering the sphere which circumscribes it cannot contain point x, which implies that this face would not have belonged to the cavity.

The second proposal is also impossible because in this case point x would be inside the circumscribed sphere at the tetrahedron; therefore, the latter should have been eliminated. As a result this face would not have belonged to the cavity.

Finally, only point x can be on the circle circumscribed to the face.

3.2.2. *Stage-2: calculation of the volume measurement common to $\acute{c}_x$ and c_v*

Once point x is inserted in the diagram, it is necessary for us to calculate the volume measurement at cell $\acute{c}_x$ associated with point x and cell c_v associated with each node v the neighbor of x (in the new diagram) before the insertion of x; a volume which we term in the following as the polyhedron P^{int}. To calculate this measurement we can proceed in several ways.

3.2.2.1. *By the recursive Lasserre algorithm*

Like Sambridge *et al.* [SAM 96, BRA 95b], Cueto *et al.* [ALF 07] proposed to use the recursive Lasserre algorithm, which needs, for the calculation of the volume measurement of a convex polyhedron in $\mathbb{R}^n$, only for all of the half-planes to define it, thus being freed from an explicit topological description of the latter.

In the case of a convex field, the Voronoi cells of the nodes of the latter are convex; consequentially, the polyhedron P^{int} being convex (intersection of two convex polyhedrons) can then be described as a volume delimited by a finite half-plane set (H-representations). For a nearby node v_i given by x, this set consists of all the faces of c_{v_i}, nodal cell v_i before insertion of item x, and the face common to nodes v_i and x (Figure 3.6). The polyhedron P^{int} being described by one (H-representations), we can consequently calculate the measurement of its volume by using the recursive Lasserre algorithm.

That is to say for a convex polyhedron P in $\mathbb{R}^n$, delimited by m half-planes, each one of the latter can be represented by an inequation, which can be translated into a matric form by the inequation

$$\boldsymbol{A} \cdot \boldsymbol{x} \le \boldsymbol{b}, \qquad [3.2]$$

where $\boldsymbol{A}$ is a matrix $m \times n$, $\boldsymbol{x}$ is the column of n components of item x, and $\boldsymbol{b}$ is a column of m terms.

The cubic measure of P, noted as $V(n, \boldsymbol{A}, \boldsymbol{b})$, can then be calculated by the following formula:

$$V(n, \boldsymbol{A}, \boldsymbol{b}) = \frac{1}{n} \sum_{i=1}^{m} \frac{b_i}{\|\boldsymbol{a}_i\|} V_i(n - 1, \boldsymbol{A}, \boldsymbol{b}) \qquad [3.3]$$

with the $V_i(n - 1, \boldsymbol{A}, \boldsymbol{b})$ volume measurement in R^{n-1} of the face f_i:

$$f_i = \left\{ x \mid \boldsymbol{a}_i^T \cdot \boldsymbol{x} = b_i, \ \boldsymbol{A} \cdot \boldsymbol{x} \le \boldsymbol{b} \right\}.$$

NOTE:–

Equation [3.3] can be put in the known form (in 2D or 3D):

$$V(P) = \frac{1}{n} \sum_{i=1}^{m} \text{Dist}(o, f_i) \cdot V(f_i), \qquad [3.4]$$

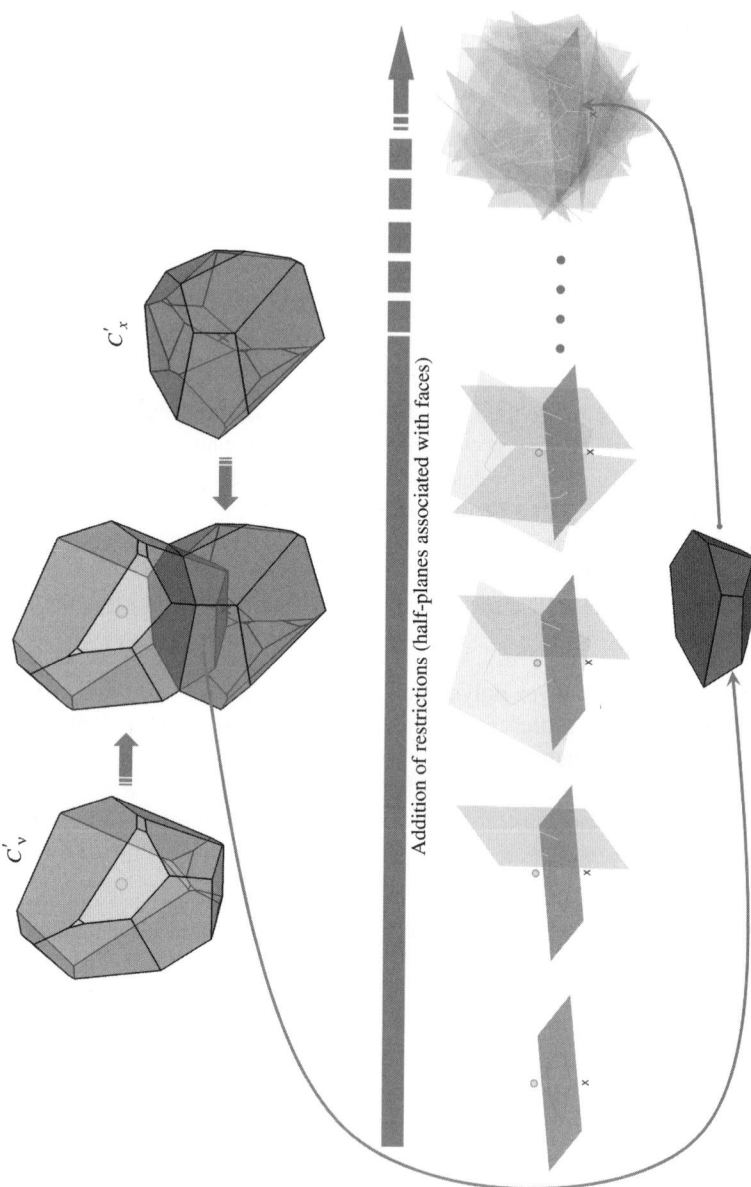

Figure 3.6. *Stage-2: volume measurement calculation of the polyhedron P^{int} by Lasserre*

where o is an unspecified point and $\text{Dist}(o, f_i)$ is the distance designated at o with the face f_i ($\vec{op} \cdot \vec{n}$, with p a point of f_i, and $\vec{n}$ normal external unit of f_i).

$V_i(n - 1, \boldsymbol{A}, \boldsymbol{b})$ cannot be calculated directly because we do not know the polyhedron nodes, which define the face f_i explicitly. Lasserre thus proposes to project the latter in a subspace of dimension $n - 1$.

The face f_i is defined by fixing the ith stress in inequation [3.2] ($\boldsymbol{a}_i^T \cdot \boldsymbol{x} = b_i$). We choose $a_{ij} \neq 0$ to substitute the variable $x_j = b_i - \sum_{k=1, k \neq j}^n \frac{a_{ik} \cdot x_k}{a_{ij}}$ in the $m - 1$ of other inequations, to lead to the system:

$$\tilde{\boldsymbol{A}} \cdot \tilde{\boldsymbol{x}} \leq \tilde{\boldsymbol{b}}, \tag{3.5}$$

which defines the face $\tilde{f}_{ij}$, orthogonal projection of the face f_i on a hyperplane of normal $\vec{e}_j$.

$\tilde{\boldsymbol{A}}$ is a matrix $\tilde{m} \times \tilde{n}$ and $\tilde{\boldsymbol{b}}$ is a column of $\tilde{m}$ terms, where $\tilde{m} = m - 1$ and $\tilde{n} = n - 1$. $\tilde{\boldsymbol{x}}$ is column ($\tilde{n}$ terms) components of x to which x_j was withdrawn (the substituted variable).

It is then possible to write

$$V_i(n - 1, \boldsymbol{A}, \boldsymbol{b}) = \frac{1}{\lambda} \cdot V_i(\tilde{n}, \tilde{\boldsymbol{A}}, \tilde{\boldsymbol{b}}), \tag{3.6}$$

where λ is the relationship between the volume measurement of the face f_i and its orthogonal projection $\tilde{f}_{ij}$:

$$\vec{n}(f_i) \cdot \vec{e}_j = \frac{a_{ij}}{\|\boldsymbol{a}_i\|},$$

$\vec{n}(f_i)$ is the unit normal with the face f_i.

To illustrate this see Figure 3.7, which represents a 3D case.

By combining equations [3.6] and [3.3], we obtain the recursive Lasserre algorithm:

$$V(n, \boldsymbol{A}, \boldsymbol{b}) = \frac{1}{n} \sum_{i=1}^m \frac{b_i}{a_{ij}} V_i(\tilde{n}, \tilde{\boldsymbol{A}}, \tilde{\boldsymbol{b}}). \tag{3.7}$$

After $n - 1$ substitutions, at the end of each branch of the recursive diagram, we are led to a field in R, which is represented by a system of $m - n + 1$ inequations,

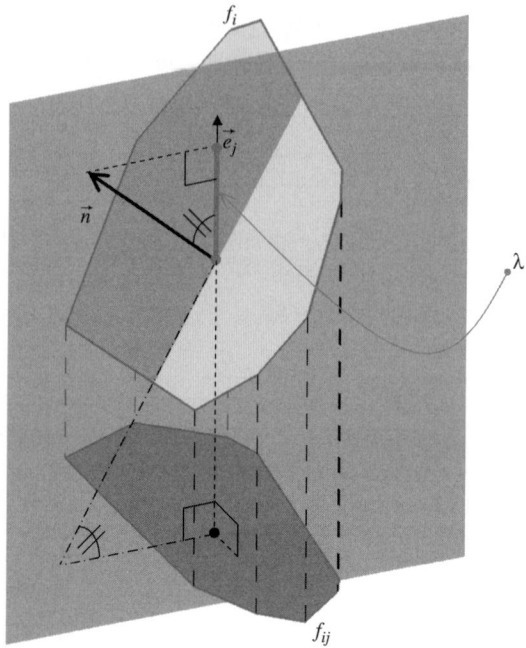

Figure 3.7. *Orthogonal projection of face f_i on a normal plane $\vec{e_j}$*

and which is of the form $\boldsymbol{\alpha} \cdot x_l \leq \boldsymbol{\beta}$, with detailed columns $\boldsymbol{\alpha}$ and $\boldsymbol{\beta}$ the remaining variables. The volume measurement of such a field is equal to

$$\max\left\{0, \left(\min\left\{\frac{\alpha_i^{\text{positive}}}{\beta_i}\right\} - \max\left\{\frac{\alpha_i^{\text{negative}}}{\beta_i}\right\}\right)\right\}. \qquad [3.8]$$

IMPORTANT NOTE. This method can only be applied to convex polyhedrons P^{int}. For non-convex fields, certain cells can be non-convex, which prohibits the use of the recursive Lasserre algorithm in this context.

3.2.2.2. *By means of a complementary volume*

The volume measurement of the polyhedron P^{int} for a given neighbor v_i can also be defined as being the measurement of the variation of the volume of this node cell after insertion of point x (Figure 3.8).

Let $N_V(x)$ be the set of nodes neighboring x and $S(N_V(x))$ the set of the vertices of the node cells of set $N_V(x)$ (after insertion of point x).

Thanks to $S(N_V(x))$, it is possible to rebuild the topology (cells, faces, edges, and vertices) of the cells of set $N_V(x)$, and also of x. By simple subtraction, we can

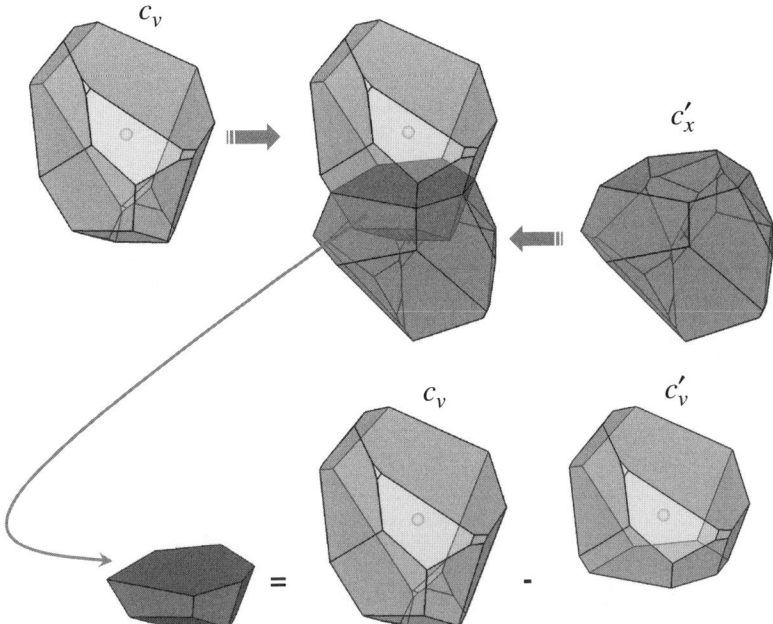

Figure 3.8. *Stage-2: Calculation of the volume measurement of the polyhedron P^{int} by means of a complementary volume*

thus deduce from this the variation of the volume measurements of the cells associated with the nodes neighboring point x.

This method can be applied only if the cell of each neighboring node is non-infinite. If this is not the case, it is impossible to calculate the measurement of its volume, but considering that we have the need only for the measurement of the variation of volumes of the cells after insertion of point x, and not of their volume, this impossibility can be circumvented by truncating the infinite cells (addition of faces).

The rebuilding of the topology of the cells of set $N_V(x)$ and that of x is necessary to calculate their volume by using equation [3.4]. To reduce the calculation times, we can calculate these volumes in another way by using a decomposition in elementary volumes associated with each edge of each cell (Figure 3.9).

These elementary volumes are constructed as follows:

Take a the considered edge s_1, its two vertexes s_2, and n_1, n_2, n_3 its three generating nodes. The contribution of an edge a to the cell volume having n_1 as a node is formed by the two tetrahedrons T_1 and T_2, with:

– T_1 tetrahedron having n_1, S_1, S_2, n_{12} for vertices (mediums of the segment (n_1, n_2)),

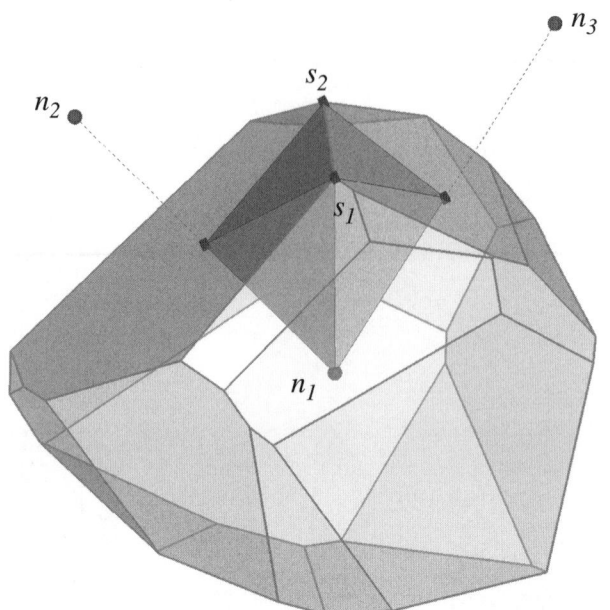

Figure 3.9. *Contribution of volume associated with an edge with cell volume*

– T_2 tetrahedron having n_1, S_2, S_1, n_{13} for vertices (mediums of the segment (n_1, n_3)).

It is, therefore, sufficient to examine all the edges formed by the vertices of the unit $S(N_V(x))$, and to add the elementary volume measurement associated with each node cell (included in the set $N_V(x)$) of the edge considered.

3.2.2.3. *By topological approach based on the CVD*

The polyhedron P^{int} generated by the intersection of $\acute{c}_x$ cell of point x, and c_{v_i} cell of a neighboring node v_i before insertion of node x, is specified. It has a particular property owing to which it is possible to rebuild its topology (faces, edges, and vertexes) while being based only on connectivities of the CVD. No actual operation of geometrical intersection is necessary.

Indeed, in general, if we take two volumes (A and B) that interpenetrate, we can break up the surface of the intersection volume into two quite distinct parts: ΓA_B (part of surface ΓA of volume A which is inside volume B) and ΓB_A (part of surface ΓB of the volume B which is inside volume A). ΓA_B and ΓB_A have a curve S in common, at the intersection of ΓA and ΓB (Figure 3.10). In our case, if cell $\acute{c}_x$ is considered as volume A, and c_{v_i} as volume B, ΓA_B is reduced exactly to $f_i \acute{c}_x$ face of cell $\acute{c}_x$ having node v_i as the second node.

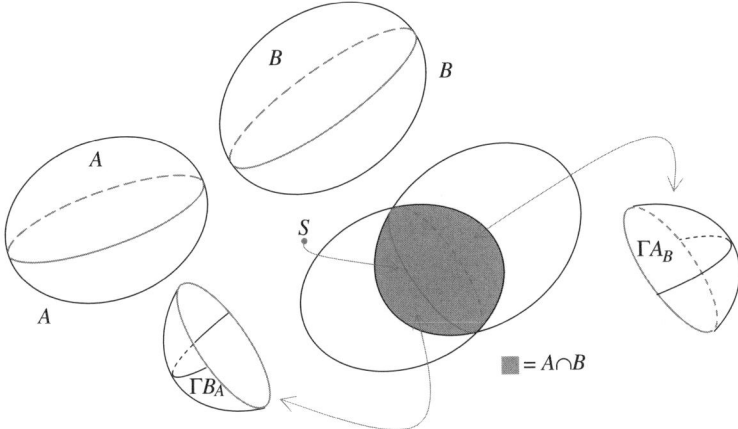

Figure 3.10. *Volumes A and B which interpenetrate*

ΓB_A can be divided in two sets: F_{trq} the set of faces of cell c_{v_i} truncated by the edges of $f_i \acute{c}_x$, and possibly $F_{\overline{trq}}$ the set of the faces of cell c_{v_i} being completely inside $\acute{c}_x$ (Figure 3.11)

The algorithm is thus composed of two stages: the first builds the F_{trq} unit, and the second builds the $F_{\overline{trq}}$ unit, and this is for each node v_i the neighbor of node x. With the volume constructed, its measurement is calculated with Equation [3.4]. For what follows, see Figure 3.12.

Let v_i be the neighbor considered, and c_{v_i} its associated cell. And let $f_i \acute{c}_x$ be the cell face of point x having v_i as one of these two nodes.

Each edge $a_j f_i \acute{c}_x$ of $f_i \acute{c}_x$ truncates a face $f_j c_{v_i}$ of c_{v_i}, a face having as one of its two nodes: the node v_{ij} and node of the edge $a_j f_i \acute{c}_x$ different from x and v_i. The two vertices of the edge $a_j f_i \acute{c}_x$ truncate two edges of $f_j c_{v_i}$, $a_k f_j c_{v_i}$, and $a_l f_j c_{v_i}$, edges having nodes v_{ij}^- and v_{ij}^+, respectively, different edge nodes of x and v_i, edges $a_j^- f_i \acute{c}_x$ and $a_j^+ f_i \acute{c}_x$, edges of $f_i \acute{c}_x$ which precedes and which succeeds $a_j f_i \acute{c}_x$. Let A be the sub-assembly of edges of $f_j c_{v_i}$, which succeeds the edge $a_k f_j c_{v_i}$ up to $a_l f_j c_{v_i}$.

The list of edges of the truncated face $f_j c_{v_i}$ is made up of truncated $a_k f_j c_{v_i}$, followed by the sub-assembly of edges A, the truncated edge $a_l f_j c_{v_i}$, and finally of the edge $(-)a_j f_i \acute{c}_x$.

With each edge a of A, couples are built (*edge, node*), where "edge" is A, and "node" v_i, which are then added to a list of couples L.

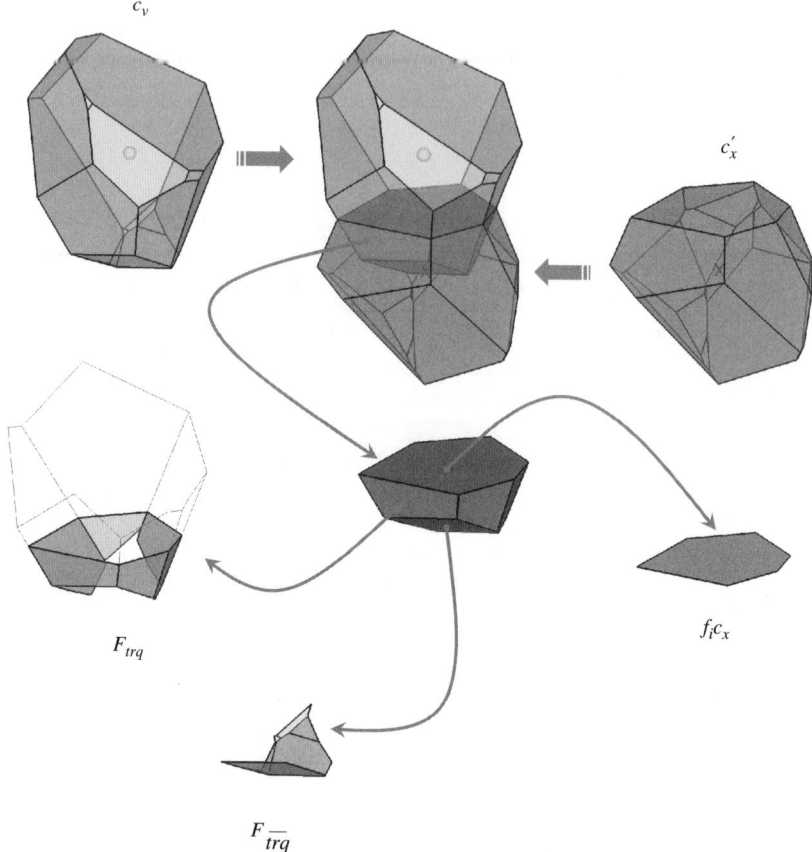

Figure 3.11. *Stage-2: calculation of the volume measurement of the*
polyhedron P^{int} by topological approach based on the CVD

Once the set F_{trq} is built, we find the faces of the set $F_{\overline{\mathrm{trq}}}$ adjacent to the faces of F_{trq} owing to the edges of the list of couples L, which can be explained as follows:

– as long as the list L is non-empty, we move the couple c to the back. Let f be the face of c_{v_i} having node v, the node of the edge of the couple c different from v_i and the node of the couple c. f belongs to the set $F_{\overline{\mathrm{trq}}}$. It is thus added to the latter. For each edge a of f different from the edge of the couple c, we verify if it is present in a couple L, if so, this couple is removed. If not, we add to L the couple (a, v).

The topological approach has the following advantages compared to the two preceding approaches:

– saving calculation time (approximately by a factor of about 10 compared with Lasserre, as is shown later),

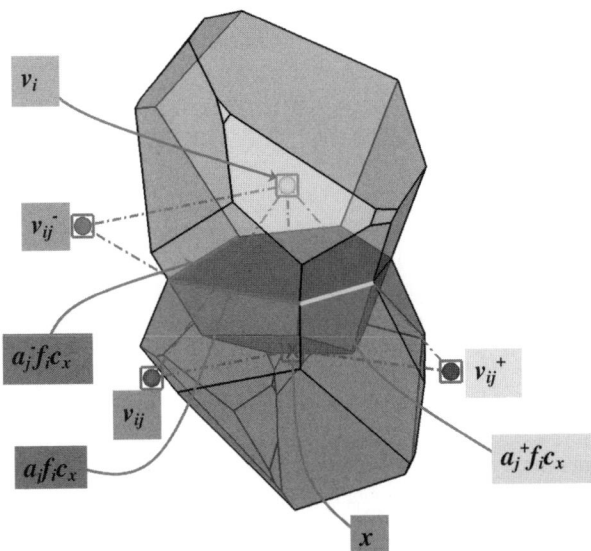

Figure 3.12. *Identification of the entities used in the topological approach based on the CVD*

– indifferent to whether the neighboring cells are infinite or not (it does not allow us to calculate the variation of the volume measurement of a neighboring cell but the volume measurement of the polyhedron intersection P^{int}),

– indifferent to whether the neighboring cells, and in particular that of point x, are convex or not.

3.2.2.4. *By topological approach based on the Constrained Delaunay tetrahedization (CDT)*

Boissonnat *et al.* [BOI 02] proposed an algorithm entirely based on the Delaunay tetrahedization (which is also applied to a CDT). The same decomposition as the previous polyhedron P^{int} is made. Let F_{trq} be the set of all the truncated faces of all the polyhedrons P^{int} associated with all the neighbors of x, and let $F_{\overline{\text{trq}}}$ be the set of all the non-truncated faces of all the polyhedrons P^{int} associated with all the neighbors of x. The construction of these two units is done in the way as described hereafter.

Let C be the cavity generated by the insertion of point x in the CDT (group of tetrahedrons whose circumscribed sphere contains point x).

Let A be the set of axes of tetrahedron C. A breaks down into two sets: A_{int} sub-assembly of the internal edges to C, and A_{ext} sub-assembly of the external edges to C (edge being located on a face – of tetrahedron – external to C).

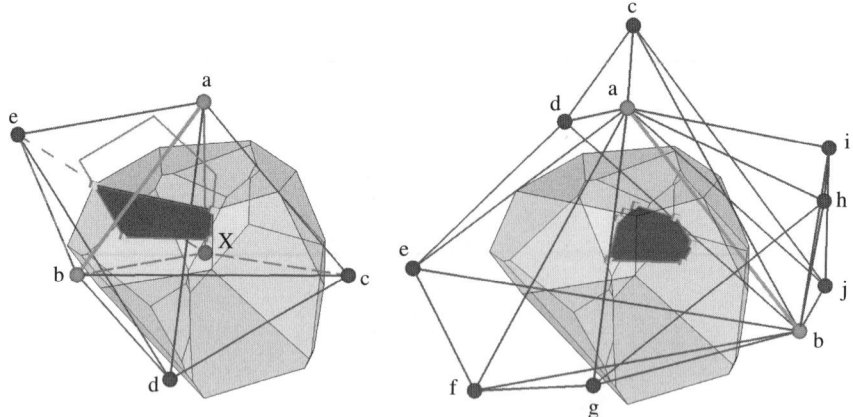

Figure 3.13. *Stage-2: calculation of the volume measurement of the polyhedrons P^{int} by topological approach based on the CDT*

Each external edge a_{ext} (edge $a - b$ Figure 3.13 on the left) of A_{ext} has for a dual a face f_{trq} (red face) of F_{trq}. The vertices associated with the tetrahedrons of C have a_{ext} for an edge constituting part of the list of the vertices of f_{trq}. To complete this list, it is necessary to add at its ends, the two new vertices associated with the tetrahedrons having point x for nodes and the nodes with the two external faces of the cavity which have as an edge a_{ext} (faces a–b–c and b–a–e).

Each edge a_{int} (edge a–b, Figure 3.13 on the right) of A_{int} has for a dual a face $f_{\overline{\text{trq}}}$ (red face) of $F_{\overline{\text{trq}}}$, this face has the vertices associated with the tetrahedrons, with C having a_{int} for an edge.

Once these two sets are built, it is sufficient to add each face of these two sets to the polyhedron P^{int} associated with each of the two nodes of the tetrahedron edge dual to this face. We finally add to each polyhedron P^{int} associated with a given node close to x, the face of the cell of x having this neighbor as one of its two nodal faces.

In addition to presenting the same advantages as the topological algorithm based on the CVD, and dividing the calculation times of the latter by a factor of 2, this algorithm is entirely based on the CDT, the construction of the CVD, and is thus ineffectual.

3.2.2.5. *Using the Watson algorithm*

Preliminary: Calculation of the volume measurement of a polyhedron by decomposition in Lawrence [LAW 91].

Let us take a polyhedron P, and π a hyperplane that is not parallel to any face of P. The volume measurement of the polyhedron P can then be calculated by algebraically summing the cubic measurements of simplexes $\mathrm{spx}_s^{\mathrm{Law}}$ which we can build by taking for each vertex s from P, in addition to the hyperplane π, the hyperplanes associated with the faces with P having s for a vertex. Each of these hyperplanes divides the space into two signed parts, the positive part being that including P. The sign of the volume measure of a simplex $\mathrm{spx}_s^{\mathrm{Law}}$ is the sign of the part of the space in which it is located, which is the product of the signs of this part of the space compared to each hyperplane associated with $\mathrm{spx}_s^{\mathrm{Law}}$ (the hyperplane not being taken into account).

In Figure 3.14, an example of this in 2D is illustrated. Polyhedron P having vertices a, b, c, d is broken down into triangles, and this while taking for each vertex s from P, the straight lines associated with the edges of P which have s for vertices, and the straight line π, which gives us:

- for $a \to$ straight lines (da), (ab), $\pi \to$ triangle (p_{da}, a, p_{ab}), sign $+ = -_{(da)} \times -_{(ab)}$

- for $b \to$ straight lines (ab), (bc), $\pi \to$ triangle (p_{ab}, b, p_{bc}), sign $- = -_{(ab)} \times +_{(bc)}$

- for $c \to$ straight lines (bc), (cd), $\pi \to$ triangle (p_{bc}, c, p_{cd}), sign $+ = +_{(bc)} \times +_{(cd)}$

- for $d \to$ straight lines (cd), (da), $\pi \to$ triangle (p_{cd}, d, p_{da}), sign $- = +_{(cd)} \times -_{(da)}$

Watson [WAT 01] proposes using the decomposition of the Lawrence polyhedron for the calculation of the volume measurement of the polyhedrons intersections P^{int}, which it modified by taking π as a plane idem, a plane idem passing through one of the faces of P^{int}. The volume measurement of the polyhedron P^{int} is obtained, then, by adding together algebraically volume measurements of the simplexes $\mathrm{spx}_s^{\mathrm{Law}}$ associated with the vertices not being located on π, $S_{\notin \pi}$. The face of P^{int} through which the plane π passes is not randomly selected; it is the face of P^{int} having (with the Voronoi direction) point x for a node, point of evaluation of the interpolation, and neighboring node v from x attached to the considered polyhedron intersection P^{int}. In other words, it is the median plane between point x and the neighboring node v.

The hyperplanes associated with faces P^{int} with $s_{\notin \pi}$ of $S_{\notin \pi}$ as vertices are median planes associated with the two nodes of each edge of the dual Delaunay simplex with $s_{\notin \pi}$ having v for a node.

Vertices of the simplex $\mathrm{spx}_{s_{\notin \pi}}^{\mathrm{Law}}$ are thus:

- the vertex of Voronoi $s_{\notin \pi}$,

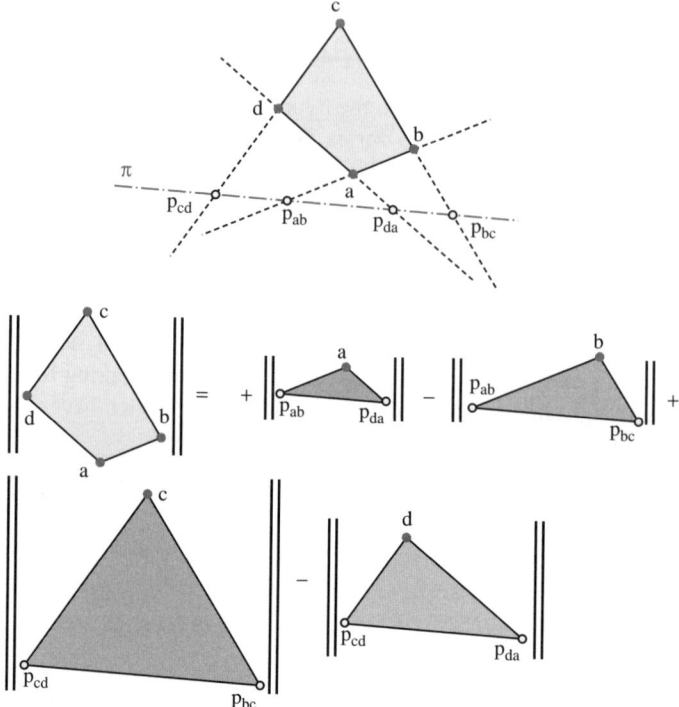

Figure 3.14. *Lawrence decomposition in 2D*

– centers of the circumscribed hyperspheres with simplexes having item x for vertices and the nodes of each face of the simplex of dual Delaunay in the vertex of Voronoi $s \notin \pi$, with node v as a node.

To illustrate this, let us take the 2D example given in Figure 3.15. The polyhedron (a, b, c, d, e, f) is the polyhedron intersection P^{int} associated with a neighboring node v of point x. Each vertex of the polyhedron P^{int} not being on the line π (vertices c, d, e, f) is a Voronoi vertex of the cavity generated by the insertion of point x, having v for a node (in the Voronoi sense). In other words it is the center of a circumscribed circle with a Delaunay triangle of the cavity having v for a node. The contribution to the volume measurement of P^{int} associated with vertex $s \notin \pi$, is the triangle constructed with the two lines associated with the edges P^{int}, having $s \notin \pi$ for a vertex, and π for the straight line. Considering that the two lines passing by $s \notin \pi$ are medians of the edges of the dual Delaunay triangle $s \notin \pi$ with v for a node, and considering the line π is also a median, right median between item x and with node v, the points of intersection of the two lines passing by $s \notin \pi$ with the straight line π are, consequentially, the centers of the circumscribed circles with the triangles having point x as a node, and the two nodes of each of the two edges of the dual Delaunay triangle with $s \notin \pi$, having v for node. For example, for vertex c:

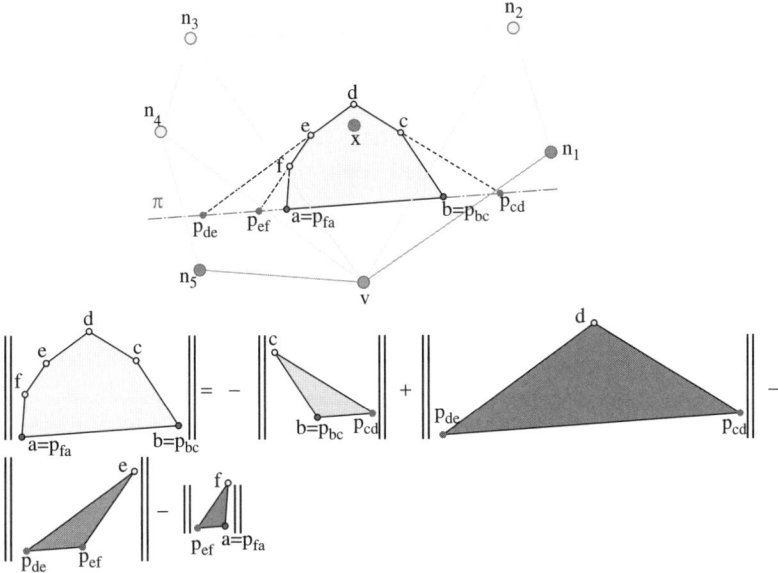

Figure 3.15. *Watson algorithm in 2D*

– the intersection of the first straight line passing by c, straight line (bc) and π, point p_{bc}, is the center of the circumscribed circle with the triangle having x as a node and v, n_1 which are the nodes of an edge of the dual Delaunay triangle in the vertex of Voronoi c having v for a node;

– the intersection of the second straight line passing through this vertex, straight line (cd) and π, point p_{bc}, is the center of the circumscribed circle with the triangle having x as a node, and v, n_2 which are also the nodes of an edge of the dual Delaunay triangle in the vertex of Voronoi c having v for a node.

This algorithm is currently the most effective one for the calculation of the Sibson shape functions. Its simplicity and its speed are due to the fact that it does not build the intersection polyhedrons P^{int} (faces, edges, vertices) for calculating the measurement of their volumes. It does it directly by looping on the simplexes of the cavity, each of the latter contributes a share to the volume measurement of each polyhedron P^{int} attached to the nodes of this simplex. It should, however, be noted that the Watson algorithm is not applicable in points of evaluation of interpolation x being located on the face of a Delaunay tetrahedron. Indeed, in this case, this face will definitely be the face of a tetrahedron of the cavity. One of the simplexes $\text{spx}^{\text{Law}}_{s \notin \pi}$ will thus have the orthocenter of a simplex as a vertex, having for a vertex the nodes of this face and the point x, which is unspecified, because the vertices of the latter are coplanar.

3.2.3. *Comparative test of the various algorithms*

The five methods of calculating Sibson-type shape functions were programmed (languages C, C++). For the calculation method based on the recursive Lasserre algorithm, we used the Vinci code [ENG 03]. In the latter, an enhanced version is implemented using the algorithm r-Las [BUE 00].

For this comparative test we used the following computer:
– OS: Microsoft Windows XP,
– compiler: Microsoft® Incremental Linker Version 7.10.3077,
– processor: Mobile Intel Pentium M 745–1800 MHz,
– RAM: 1024 Mb–PC2700 DDR SDRAM.

We took a unit cube of dimension $(1 \times 1 \times 1)$ with 10,000 nodes distributed randomly inside. The shape functions are evaluated with the barycenters of each Delaunay tetrahedron. This avoids entering the time taken by the localization of the point where we wish to evaluate the shape function in the tetrahedization. Only the points for which the five methods are applicable were used for the tests. The points were isolated where at least one of the cells of the neighboring nodes at the point of evaluation is infinite (calculation method by non-applicable complementary volume). The times given in Tables 3.1 and 3.2 are average times, corresponding to 1,000 evaluations. The total calculation time is divided into two parts:

– T_{cavity}: time taken for the construction of the cavity, evidently constant for the five algorithms tested.

– $T_{\text{cal}-ff}$: time taken by the calculation of the shape functions once the cavity is constructed.

Algorithm	Lasserre	Vol-comp	Topo-CVD	Topo-CDT	Watson
T_{cavity} (m.s)	35	35	35	35	35
$T_{\text{cal-ff}}$(m.s)	4136	544	413	192	51
$\frac{T_{\text{cal-ff}}\,\text{Alg}}{T_{\text{cal-ff}}\,\text{Alg-Wat}}$	82	11	8	4	1

Table 3.1. *Comparision of the calculation algorithms of Sibson shape functions*

As these measurements show, the Watson algorithm is the most powerful.

To locate these calculation times compared with times that could take other shape functions, we carried out the same calculation by this time, taking the Laplace natural element shape functions and linear finite elements.

f.f	Sibson(*Watson*)	Laplace	Linear EF
T(m.s)	87	150	3
$\dfrac{T}{T(\text{linear EFF})}$	33	56	1

Table 3.2. *Comparison of the calculation times for various shape functions*

The Watson algorithm thus seems to be the most suitable for the calculation of the Sibson-type shape functions in the case of the CNEM. As we have seen, this algorithm does not work for points of evaluation located on and close to a Delaunay face. For these points, the topological algorithm based on the CDT is recommended. In practice, the failure rate of the Watson algorithm for evaluation points positioned in a random way is in the order of 1/1,000.

3.3. Numerical integration

Numerical integration is one of the critical points conditioning the convergence of natural neighbor Galerkin methods. Normally, the variational formulation is integrated on the Delaunay triangles using a Hammer quadrature. For example, we know that this type of integration is a source of numerical errors [SUK 98b, CUE 02], associated with the non-polynomial character of the functions to integrate, as well as the fact that integration domains do not coincide with shape function supports. These difficulties are not specific to the NEM. They are also encountered in the majority of meshless methods (see, for example, [DOL 99]).

Two approaches were analyzed in our work: one based on the decomposition of shape function supports in much simpler forms (section 3.3.1) and the other based on a stabilized nodal formulation (section 3.3.2).

3.3.1. *Decomposition of shape function supports*

A first possibility consists of carrying out integration on intersections of shape function supports. This approach was also used by Atluri in [ATL 99] and [ATL 00] for meshless methods based on moving least-squares approximations.

In the NEM, the support of each shape function results from the union of circumscribed circles with the triangles containing the node considered. An example is illustrated in Figure 3.16. As we can see, the support and the intersection of two supports can be broken up into a certain number of triangles and circular segments. In the triangles various numerical quadratures (Gauss and Hammer) were used. For circular segments, we established a geometrical transformation to map them on the unit square.

The integration of the variational formulation on the intersection of shape function supports results in

$$\sum_I \sum_J \int_{\Omega_{IJ}} \nabla^s \delta \boldsymbol{v} : \mathbf{C} : \nabla^s \boldsymbol{u} d\Omega$$

$$= \sum_I \sum_J \int_{\Omega_{IJ}} \delta \boldsymbol{v} \cdot \boldsymbol{b} d\Omega + \int_{\Gamma_t} \delta \boldsymbol{v} \cdot \bar{\boldsymbol{t}} d\Gamma \quad \forall \delta \boldsymbol{v} \in \mathcal{V},$$

[3.9]

where Ω_{IJ} is the part of the domain Ω: $\Omega_{IJ} = \omega_I \cap \omega_J \cap \Omega$, with $\omega_I = \mathrm{supp}(n_I)$.

Using this decomposition, the integration must be evaluated on the different triangles and circular segments. The geometrical transformation for circular segments is written in the form given by equations [3.10] and [3.11] (Figure 3.17).

$$x = \frac{1}{2}\left[\frac{1}{2}(1-\eta)\left(x_1(1-\xi) + \frac{1}{2}x_2(1+\xi) + (1+\eta)\right)\right.$$
$$\left. \times \left(C_x + R\cos\left(\frac{1}{2}(1-\xi)\beta + (1+\xi)\alpha\right)\right)\right],$$

[3.10]

$$y = \frac{1}{2}\left[\frac{1}{2}(1-\eta)\left(y_1(1-\xi) + \frac{1}{2}y_2(1+\xi) + (1+\eta)\right)\right.$$
$$\left. \times \left(C_y + R\sin\left(\frac{1}{2}(1-\xi)\beta + (1+\xi)\alpha\right)\right)\right].$$

[3.11]

For this transformation, the position in the circular segment of 5×5 Gaussian points of integration defined on a unit square is illustrated in Figure 3.18.

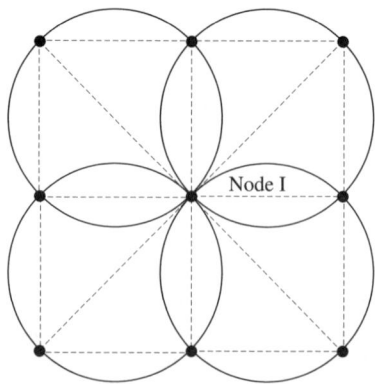

Figure 3.16. *Support of the shape function associated with node I*

3.3.2. *Stabilized nodal integration*

Recently, Chen [CHE 01] proposed a stabilized nodal quadrature applied to moving least-squares-type methods. This technique is based on a smoothing of strain, in order to define nodal strain:

$$\tilde{\varepsilon}_{ij}^h(\boldsymbol{x}_I) = \int_{\Omega} \varepsilon_{ij}(\boldsymbol{x})\Phi(\boldsymbol{x}; \boldsymbol{x} - \boldsymbol{x}_I)\mathrm{d}\Omega, \qquad [3.12]$$

where Φ is the smoothing function. In [CHE 01], this function is defined by

$$\Phi(\boldsymbol{x}; \boldsymbol{x} - \boldsymbol{x}_I) = \begin{cases} \dfrac{1}{A_I} & \text{if } \boldsymbol{x} \in \Omega_I \\ 0 & \text{otherwise} \end{cases}, \qquad [3.13]$$

where Ω_I is the Voronoi cell associated with node I and A_I represents its area. With this definition, the smoothed strain is given by

$$\tilde{\varepsilon}_{ij}^h(\boldsymbol{x}_I) = \frac{1}{2A_I} \int_{\Omega} \left(\frac{\partial u_i^h}{\partial x_j} + \frac{\partial u_j^h}{\partial x_i} \right) \mathrm{d}\Omega, \qquad [3.14]$$

and if the divergence theorem is now applied, then

$$\tilde{\varepsilon}_{ij}^h = \frac{1}{2A_I} \int_{\Gamma_I} \left(u_i^h n_j + u_j^h n_i \right) \mathrm{d}\Gamma, \qquad [3.15]$$

where Γ_I is the boundary of the Voronoi cell associated with node I.

This method of nodal quadrature produced excellent results when it was applied within the framework of RKPM formulations [CHE 01]. Also, it seems to be especially well adapted to the natural elements because the main geometrical entities

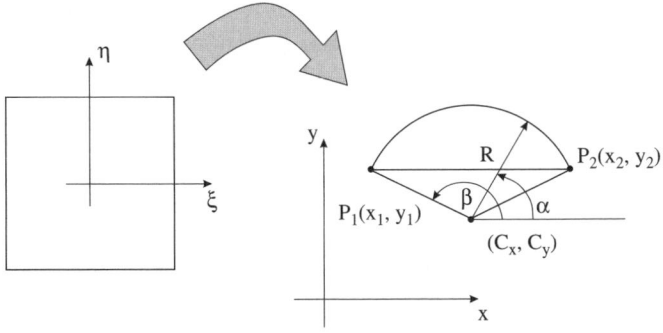

Figure 3.17. *Geometrical transformation suggested for a circular segment*

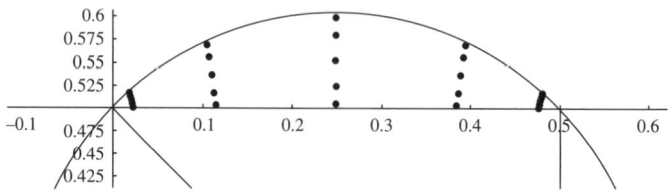

Figure 3.18. *Transformation associated with* 5×5 *Gaussian points*

intervening in the smoothing (Voronoi cells) are at the base of the definition of natural element shape functions. It should be noted that its application does not require additional calculations.

We report that the application of the divergence theorem makes it possible to avoid the calculation of shape function derivatives; only the shape functions are necessary.

A comparison between the two quadratures was carried out, for which the results are discussed in the next section.

3.3.3. *Discussion in connection with various quadratures*

To compare the two quadratures presented above, we refer to the criterion of the *patch test* introduced in [IRO 72]. Even if this criterion was initially introduced to evaluate the convergence of non-conforming formulations, it was also used to study the precision of numerical quadratures used within the framework of meshless formulations [DOL 99, SUK 98b]. The step to be followed is the imposition of a linear field on the boundary of a domain defined by several elements. When the approximation has linear consistency, a linear field should also be obtained inside the domain. If the solution obtained is not linear, we can infer that the numerical integration is not precise enough.

3.3.3.1. *2D patch test with a technique of decomposition of shape function supports*

In works relating to the NEM ([SUK 98b, CUE 02]), the patch test criterion was not satisfied with the machine precision in 2D and 3D. We consider the situation illustrated in Figure 3.19. The numerical results of the *patch test* are grouped in Table 3.3. We conclude that in spite of the great number of points of integration, we are far from the machine precision (10^{-16}). This behavior is similar to that obtained within the framework of the finite-sphere method ([DE 01]) or in the Moving Least Squares Petrov–Galerkin method [ATL 99], where many points of integration are necessary to obtain an acceptable precision. Indeed, starting from a certain number of points, the gains in terms of precision are negligible compared to the increase of computer cost.

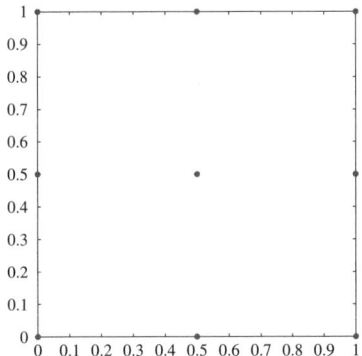

Figure 3.19. *Domain containing nine nodes where numerical integration is carried out with a technique of support decomposition*

Gaussian points (triangles)	Gaussian points (segment)	$\|e\|_{L_2}$
2×2	2×2	$2.1527 \cdot 10^{-2}$
4×4	4×4	$2.2793 \cdot 10^{-3}$
10×10	10×10	$4.1850 \cdot 10^{-4}$
20×20	20×20	$1.3944 \cdot 10^{-5}$
50×50	50×50	$6.1493 \cdot 10^{-6}$
70×70	70×70	$2.2505 \cdot 10^{-6}$
100×100	100×100	$5.6665 \cdot 10^{-7}$
150×150	150×150	$4.2297 \cdot 10^{-7}$

Table 3.3. *Error in L_2 norm for the 2D patch test*

We can conclude that the non-polynomial character of the functions to be integrated, as well as the non-polynomial character of the geometrical transformations associated with circular segments, induces significant errors of integration. This explains the interest in testing stabilized nodal integration presented previously.

3.3.3.2. *2D patch test with stabilized nodal integration*

The technique described in section 3.3.2 was applied within the framework of different *patch tests* (node clouds) in a unit square. The various node clouds are shown in Figure 3.20. As we can see, regular as well as irregular distributions were considered. In each case, only one point of integration was used on each of the internal edges of the Voronoi cells. For edges located on the boundary of the domain, we used two Gaussian points. The numerical results obtained are listed in Table 3.4.

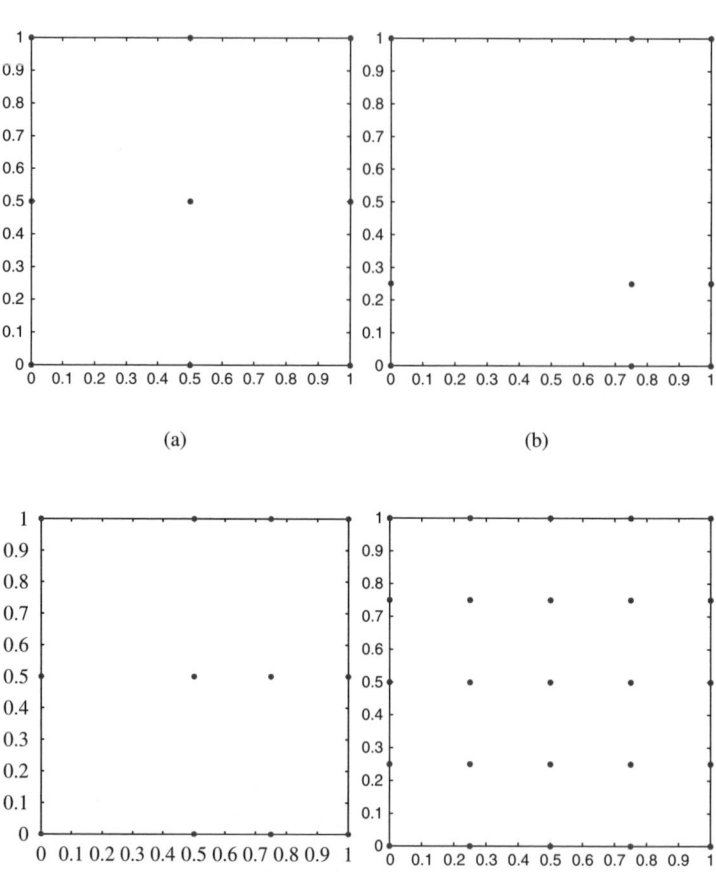

Figure 3.20. *Nodal distributions considered in the 2D patch test with stabilized nodal integration*

Two types of tests were considered: one with imposed displacements and the other with imposed tractions. In the last case, a constant stress $\sigma_y^0 = 1$ is imposed, in such a way that theoretical displacement becomes linear:

$$u_1 = \frac{\nu}{E}(1 - x_1)$$
$$u_2 = \frac{x_2}{E}$$

[3.16]

Case	Field imposed	$\|e\|_{L_2}$
a	Bilinear $\boldsymbol{u}$	$2.7208 \cdot 10^{-16}$
a	Linear $\boldsymbol{u}$	$2.4343 \cdot 10^{-16}$
b	Bilinear $\boldsymbol{u}$	$4.2407 \cdot 10^{-16}$
c	Bilinear $\boldsymbol{u}$	$2.9842 \cdot 10^{-16}$
a	$\sigma_y^0 = 1$	$7.8614 \cdot 10^{-16}$
d	$\sigma_y^0 = 1$	$9.2009 \cdot 10^{-16}$

Table 3.4. *Error in norm L_2 for the patch test with stabilized nodal integration, for the nodal distributions defined in Figure 3.20*

We have recorded much better results (very close to the machine precision) than those obtained with the technique of support decomposition in the two types of the test.

Integration on the Delaunay triangles [SUK 98b] is not sufficiently precise either, with errors in the order of 10^{-5} for nodal distributions that are very close to those considered here.

From the analysis that we have just described, we can conclude the performances from the stabilized nodal quadrature when it is used within the framework of the natural neighbor Galerkin methods.

3.3.3.3. *3D patch tests*

In this section, we finally consider the 3D case combined with a stabilized nodal integration. In [CUE 02], we proved that a Gaussian integration on the tetrahedrons does not verify the criterion of the patch test with the machine precision, as in the 2D case [SUK 98b].

As in the preceding case, here we consider a unit cube with different node clouds (Figures 3.21–3.23), and a linear displacement field is imposed on the boundary of the domain.

Integrals on the faces were calculated either with a point of integration or by breaking up these faces which are always convex. In Figure 3.24, as an example, the Voronoi cells for a cloud of nine nodes are represented. The faces intersecting the boundary of the domain are triangulated, and a numerical quadrature with three points per triangle is applied. However, for the internal faces, the results obtained with only one point of integration are sufficient.

In Table 3.5, numerical results are compared with those obtained in [CUE 02]. Distribution comprising eight nodes, where the integration is carried out with

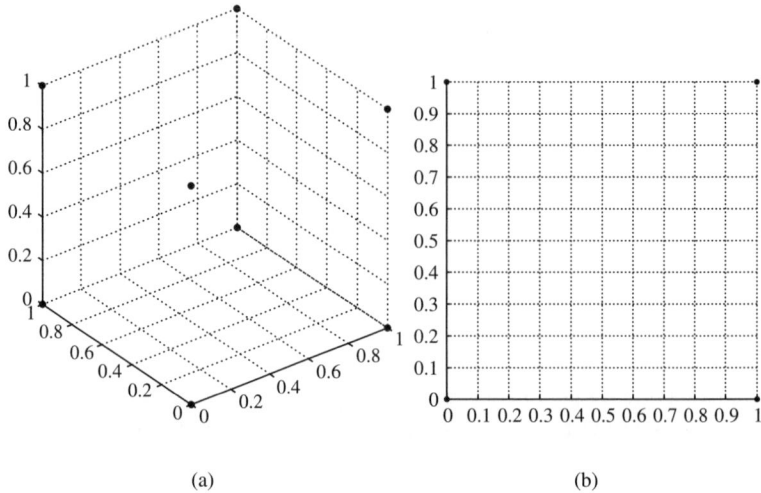

Figure 3.21. *Nodal distributions used for the application of the patch test for a cloud of eight nodes*

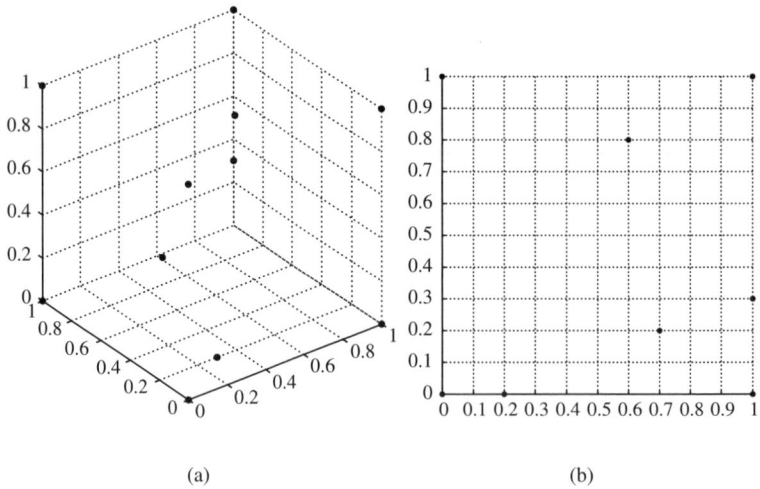

Figure 3.22. *Nodal distributions used for the application of the patch test for irregular distributions of nine nodes*

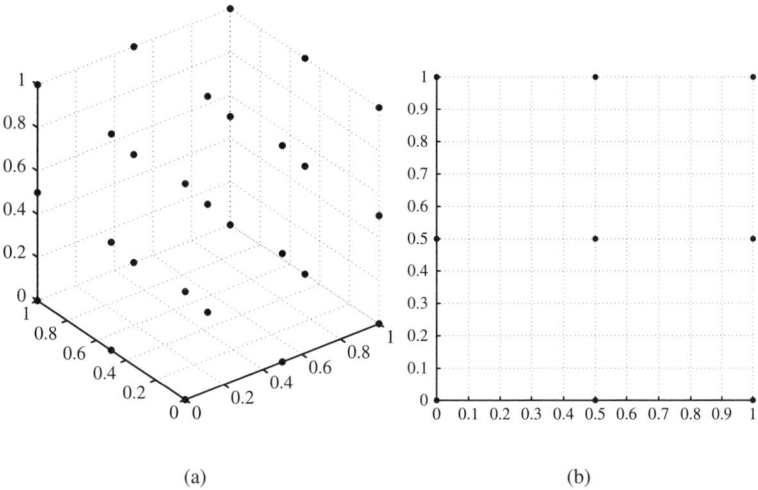

(a)   (b)

Figure 3.23. *Nodal distributions used for the application of the patch test for a distribution comprising 27 nodes*

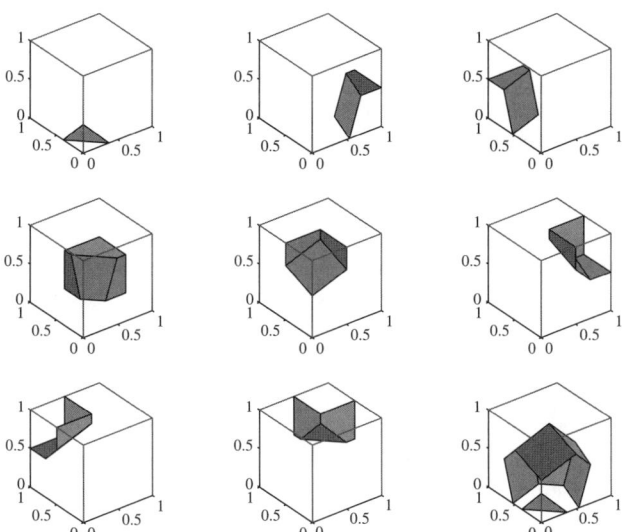

Figure 3.24. *Voronoi cells associated with each of the nine nodes – for simplicity, Voronoi faces on the boundary of the domain are not represented*

Node cloud	Number of nodes	Nature of distribution	Integration on tetrahedrons	Stabilized integration
Figure 3.21	8	Regular	1.7844×10^{-16}	4.5678×10^{-16}
Figure 3.22	9	Irregular	1.0877×10^{-2}	2.1298×10^{-16}
Figure 3.23	27	Regular	2.4660×10^{-3}	1.5796×10^{-16}

Table 3.5. L_2 *error for the 3D patch test*

four points in each tetrahedron, is a symmetrical case where the errors are compensated ([SUK 98b, CUE 02]). As soon as the nodal distribution is disturbed, the compensation disappears and the integration errors reappear.

The use of stabilized nodal integration improves the results substantially. Moreover, as already discussed, the necessary geometrical entities to put this integration into work are the same as those necessary for the calculation of the shape functions of the natural neighbor type.

3.4. NEM on an octree structure

In recent years, there has been a renewed interest in the structured mesh generation domain. For example, in 2003, Krysl *et al.* [KRY 03] proposed a hierarchical approach to the adaptive mesh in the context of the finite elements. In addition, in [KAG 03], a method based on B-spline finite elements, which considers non-conforming nodes in a quadtree-type structure, was also presented.

Certain researchers have also proposed an approach using very densely structured meshes composed of hexahedron elements of identical size [NAG 02, BEL 03]. Earlier work had already explored this channel in biomechanics where the complexity of the domains to be meshed (bone, muscles, ligaments, and so on) and the nature of the geometrical data – frequently deduced from recompositions starting from 2D images obtained, for example, by magnetic resonance imaging – generally lead to a very complex mesh generation. In [KEY 90] or [RIE 96] proposed a direct conversion of voxels (defining 3D images) into hexahedral elements. This technique leads to a completely regular finite-element mesh which has, as a direct consequence, ease of matrix assembly, (stiffness, mass, and so on) as well as a resolution of the associated systems of equations, owing to specific techniques. These techniques include, of course, the possibility of parallelization by means of Element-by-Element [HUG 87] or Row-by-Row techniques [RIE 96]. If the voxels are too small to be directly transformed into elements, it is possible to join them together by regular packages. The finite-element mesh of a femur obtained by this technique is presented in Figure 3.25.

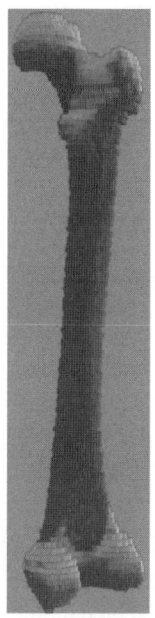

Figure 3.25. *Finite-element mesh of a human femur deduced from an image composed of voxels (courtesy of "Advanced Computer Graphics Group," Universidad de Zaragoza)*

However, these various meshing techniques present some disadvantages. As all elements have the same size – cubic hexahedrons, sometimes simply rectangles with a larger size than the other two – the resulting mesh does not correctly respect, in general, the domain boundary. This leads to solutions filled with large errors in the vicinity of the boundary and, in particular, in the interfaces between different materials. In [BEL 03, NAG 02], the extended finite element method (X-FEM) is used to control the representation of the discontinuity of the essential variables or their derivatives. The X-FEM belongs to the methods based on the partition of unity [BAB 97]. It makes it possible to take cavities into account without requiring a mesh compatible with the latter.

However, there remains an important aspect to analyze which is common to these approaches. As the mesh is highly structured, it is not possible to put adaptive mesh techniques into practice based on the use of elements of variable sizes (h-adaptivity).

The appearance of the meshless methods [BEL 94, DUA 96, SUK 98b] has opened new pathways to deal with such problems due to the absence of limitation regarding the relative positions between nodes. However, the meshless approaches are generally more costly than the finite-element approaches which prohibit them, *a priori*, from dealing with the problems of such an important size.

In [KLA 00] a method is proposed to somewhat speedup meshless calculations. It is based on the use of moving least squares functions to perform the approximation in a Galerkin framework, while interacting nodes are found in an octree-based search. In that work, however, the nodal distribution does not maintain the binary tree structure and thus no speedup is found in the element stiffness matrix calculation. In [MAC 03] the point placement is performed by employing the tree structure in the method of Finite Spheres [DE 01] in order to speedup the calculations.

Below, we introduce a method to generate highly structured numerical models taking into account the possibility of h-adaptivity. This also considers a binary tree structure, both in the neighboring nodes search and in the computation of the element stiffness matrix, which is the main advantage over previous approaches, like [MAC 03]. We consider a binary tree structure where a node is placed at each binary cell corner. Thus, it is possible to perform a fast location of neighboring nodes and, since the number of possible neighboring patterns is finite, the element stiffness matrix of each of these patterns can be analytically (in the simplest cases) or numerically determined, and stored.

Data is interpolated, in a Galerkin framework, through the use of natural neighbor interpolation [SIB 81]. Thus, the presented method can be seen as a particular instance of Natural Neighbor Galerkin methods [SUK 98b, CUE 02]. This enables us to maintain the quadtree structure of the nodes, obtaining a continuous essential field, albeit "hanging" nodes are present in the discretisation. Essential boundary conditions are imposed by the characteristic function method [BAB 03], using R-functions [RVA 95].

3.4.1. *Structure of the data*

3.4.1.1. *Description of the geometry*

The quadtree structures, or their 3D equivalent *octrees*, are largely used for geometrical modeling [MEA 82] or for the generation of finite-element meshes [YER 84]. The term *quadtree* [SAM 84] refers to a class of hierarchical data structures based on the principle of the recursive decomposition of space. The structure begins by embedding the whole domain within a square, that represents the root of the quadtree. This square is then divided into four other squares by bisection, up to the user's desired level of decomposition (see Figure 3.26).

The various types of quadtrees can be classified according to [SAM 84, 95]:

1. the type of data (in this particular instance, regions of the space);

2. the principle guiding the decomposition process and;

3. the resolution, that can be variable or not.

In our case, the decomposition process can be guided, for instance, by the curvature of the curve representing the boundary. Thus, the tree is refined, depending

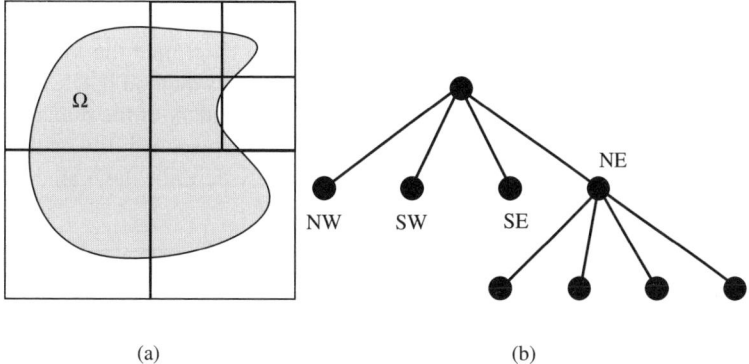

(a)   (b)

Figure 3.26. *Quadtree decomposition of an arbitrary area (a) and associated organization of the data (b)*

on the level of detail desired to represent the boundary. This is the approach followed to generate the quadtree structure of the femur head shown in Figure 3.27. Another possibility is to fix *a priori* a certain level of recursivity, for certain areas of the field, or to be guided by rules pre-established by the user according to his or her knowledge of the form of the required solution (areas with strong gradients, for example).

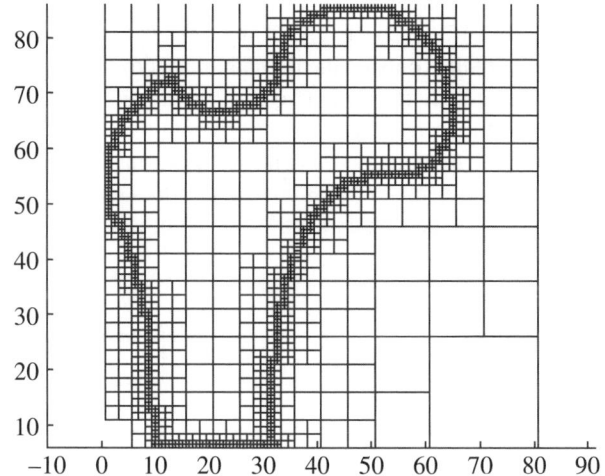

Figure 3.27. *Quadtree domain decomposition representing a human femur head*

Whenever the boundary of the domain is not explicitly defined (as in the case of data obtained from medical images, where the boundary is usually extracted by a segmentation procedure) a very convenient method to determine the intersection of the boundary with each cell is to use the *marching squares* method [LOR 87]. In this case, it is supposed that the intersection between the boundary of the domain and the cells is done by the segments of a straight line. It is the responsibility of the user to choose an appropriate quadtree refinement to accurately describe the boundary.

Another, more simple, possibility to represent the domain is, as mentioned before, trough CAD descriptions. The simplest of these descriptions is a Planar Straight Line Graph (PSLG), which is composed of line segments that approximate the boundary of the domain. In this case, the boundary is intersected with the quadtree cells and it is assumed that within each cell the boundary is a straight line in 2D or a plane in 3D, following the same marching squares patterns mentioned before. In three-dimensional cases, the STL format (stereolithography format) is also a good means to describe the geometry of the domain. Briefly, it consists of a list of triangular facets that describe a computer generated model of a solid. These facets are also enhanced with the normal. In order to construct the octree structure of the method, the triangle list must be crossed to determine its intersection with the cells.

Three types of nodes are to be considered: interior nodes, external nodes, and nodes on the boundary. A node is said to be on the boundary if an edge of the domain boundary passes through one of the cells connected to the node. A node is said to be interior, (outside, respectively) if the cells that are connected there are completely inside the domain, (outside the domain, respectively). Each cell is broken up in a recursive way up to the maximum level of recursivity, n. It has been shown [KLA 00] that it is possible to reach the neighboring information between the nodes in constant time if this information is stored on each quadrant and the difference in the level between two neighboring quadrants is controlled (does not exceed a fixed value). In the following, this difference was fixed at 1, for the preceding reason and also for other reasons given later (Figure 3.28).

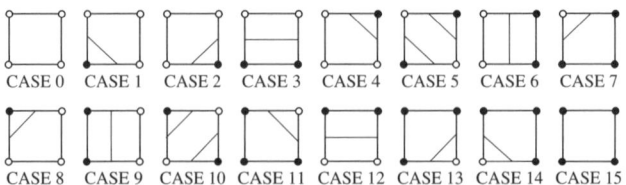

Figure 3.28. *Sixteen different marching squares cases; black dots represent the nodes inside the field and the white circles those on the outside; cases 5 and 10 are ambiguous; only one of two possibilities has been represented*

One of the main advantages of the use of the quadtrees to represent the domain is the possibility of implementing recursive processing. If we define the set Q as the set of quadrants describing a given quadtree rooted by Q_0, and assigning the Level$(Q_0) = 0$, we can define the level of a child quadrant as

$$\text{Level}(Q_i) = \text{Level}(\text{Parent}(Q_i)) + 1. \tag{3.17}$$

Thus, for example, the size of a cell is given by

$$\text{Size}(Q_i) = \frac{1}{2}\text{Size}(\text{Parent}(Q_i)). \tag{3.18}$$

In the method suggested, once the nodes of a given quadrant are identified, the following stage consists of the description of the structure to build the interpolation.

3.4.1.2. Interpolation on a quadtree

One of the attractions of the quadtree structure for the natural neighbor interpolation is the limited number of possible neighboring arrangements. In the work presented here, the proposed approach benefits from the limited number of neighboring patterns that are possible for a given two-dimensional cell, if the level difference between neighboring cells is controlled, as mentioned before. It is thus interesting to store the value of the shape functions at the points of integration rather than the elementary matrices themselves. This makes it possible, among other things, to consider non-homogeneous materials. Various possible arrangements around a quadrant surrounded by eight quadrants of the same level or quadrants of a higher level are represented (as well as the corresponding Voronoi diagram) in Figure 3.29. If the quadrant is surrounded by cells of a lower nature, similar arrangements are found. The number of values of the shape functions to be stored depends, evidently, on the type of the integration diagram selected. It should be noted that the level of the cells in the corners does not intervene on the neighboring of the central cell, which limits the number of cases to be considered.

The extension of this approach to the 3D case follows the same step. The number of arrangements is, however, appreciably larger, and it is thus the same for the number of values of the shape functions to store.

3.4.1.3. Numerical integration

Here, we propose to use a Gaussian-type integration diagram with three points of integration on each Delaunay triangle, resulting from the decomposition into quadrants earlier introduced. Other integration diagrams can be used such as stabilized nodal integration [CHE 01, GON 04a, YOO 04].

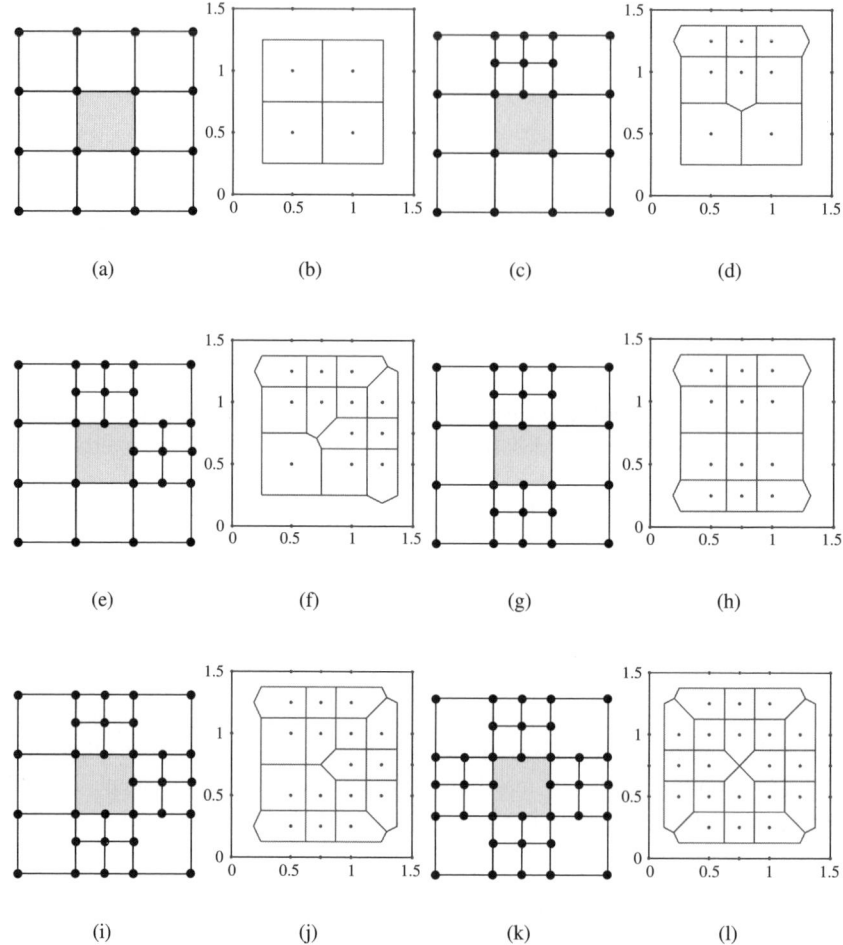

Figure 3.29. *Possible arrangements of neighborhood, and the corresponding Voronoi diagram, for a cell surrounded by others of the same level or higher level*

The nodes belonging to quadrants external to the domain are eliminated from the model. Obviously, the weak form is only integrated on the domain considered.

For interior cells – those completely within the domain – numerical integration is performed by dividing the cell into two triangles and employing three-point quadrature in them. This has been preferred to (2×2) Gauss quadrature over the cells since it is expected, in a general case, that this quadrature will better capture the influence of neighboring nodes, exterior to the cell.

Finally, boundary, polygonal-shaped, cells are split into triangles and the same three-point quadrature rule is employed.

3.4.2. *Application of the boundary conditions – interface conditions*

A crucial step in the construction of a model with the method proposed here is to enforce essential, interface and natural boundary conditions, since the structure described by the quadtree, as explained before, does not conform, in a general case, with the geometry of the domain. In order to impose essential boundary conditions we have chosen, among a wide range of possibilities, the employ of R-functions. Other possibilities include the use of penalty formulations or Lagrange multipliers [BEL 94]. To deal with piece-wise homogeneous domains, however, we have chosen to implement the method as a particular instance of the family of partition of unity methods [BAB 96, BAB 97].

3.4.2.1. *Dirichlet-type boundary conditions: use of R-functions*

Among the various methods classically used by the meshless approaches to apply the boundary conditions [BAB 03] (Lagrange multipliers, Nitsche, collocation, coupling with a finite-element band, and so on), we propose to use the characteristic function approach. For a field having a smooth boundary $\partial\Omega$, regular functions ω exist, verifying

$$\omega(x) > 0, \quad x \in \Omega, \qquad\qquad [3.19]$$

$$\omega(x) = 0, \quad x \in \partial\Omega, \qquad\qquad [3.20]$$

$$|\nabla\omega(x)| \geq \alpha > 0, \quad x \in \partial\Omega. \qquad\qquad [3.21]$$

Let $\mathcal{S}_h^\omega = \{u | u = \omega v, \ v \in \mathcal{S}_h\}$, where $\mathcal{S}_h$ represents a finite-dimensional subspace of the trial space H. Then it is immediate that $\mathcal{S}_h^\omega \subset \mathcal{V}$. Error bounds and proof of convergence of this method can be found in [BAB 03]. Completeness of this type of approximation is also proved in [RVA 00].

This method originated in the early applications of the Ritz method using global polynomials as trial functions (see [BAB 03] and associated references). A renewed interest for this step appeared later in the work of Shapiro *et al.* [SHA 99] on the R-function theory. It is clear that this approach offers hardly any interest if we cannot easily produce such functions ω. The R-functions behave as a tool box making it possible to build these functions, and are used in this work to build the characteristic functions so as to verify the Dirichlet-type boundary conditions.

The R-function is a real function whose sign is entirely defined by the sign of its arguments. Such functions define Boolean operations which help to build combinations of simple basic functions. Let us consider, for example, the following functions which behave as logical operators *and* and *or*:

$$x \wedge y \equiv (x + y - \sqrt{x^2 + y^2}), \qquad [3.22]$$

$$x \vee y \equiv (x + y + \sqrt{x^2 + y^2}). \qquad [3.23]$$

Let us consider, for example, the domain represented in Figure 3.30. The domain is defined through six inequalities of the form:

$$f_1 = y - y_1 \geq 0 \qquad [3.24]$$

$$f_2 = x - x_2 \geq 0 \qquad [3.25]$$

$$f_3 = y - y_3 \geq 0 \qquad [3.26]$$

$$\vdots$$

These inequalities can be combined so as to define the complete domain by

$$\Omega = (f_1 \vee f_2) \wedge f_3 \wedge f_4 \wedge f_5 \wedge f_6. \qquad [3.27]$$

An example of the application of this technique is shown in Figure 3.31. Dirichlet-type boundary conditions are shown in Figure 3.31(a) by black lines, and the resulting R-functions are shown in Figure 3.31(b).

Non-homogeneous boundary conditions can be written in the form

$$\mathbf{u}|_{\partial\Omega} = \boldsymbol{\varphi}_0 \qquad [3.28]$$

a function φ such as $\text{trace}(\varphi)|_{\partial\Omega} = \varphi_0$ can be added to the solution for homogeneous boundary conditions. To obtain such a function, by assuming that φ_0 is imposed in a

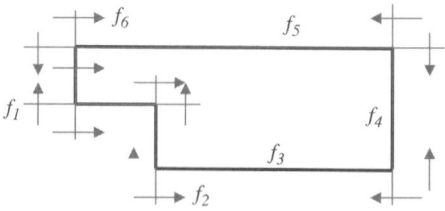

Figure 3.30. *Definition of the domain boundary given by six inequalities*

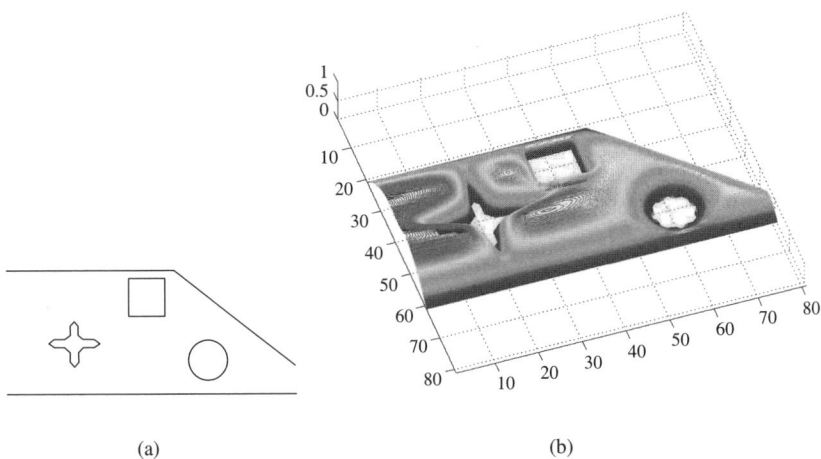

(a)   (b)

Figure 3.31. *Boundary conditions of Dirichlet (a) and associated*
R-function (b)

piece-wise manner at each portion of the PSLG $\partial\Omega_i$, a total function φ can be obtained by

$$\varphi = \frac{\sum_{i=1}^{m} \varphi_i \prod_{j=1,\ j\neq i}^{m} \omega_j}{\sum_{i=1}^{m} \prod_{j=1,\ j\neq i}^{m} \omega_j}, \qquad [3.29]$$

where φ_i represents the function prescribed on each portion of the boundary and ω_j is a R-function vanishing at the boundary portion $\partial\Omega_j$. m is the number of segments defining the boundary.

As we can see in Figure 3.31, the R-functions behave in a way similar to distance functions. The R-functions can also be used to impose conditions on the interface connecting two different materials. Through this interface, displacements are continuous, but the strains are not. This discontinuity can be added to the enrichment functions also in the form of an R-function. For example, this was done in [BEL 03] and [SUK 01a] thanks to a distance function approximated by finite-element shape functions. As the discontinuity relates only to the nodes whose support is cut by the interface, it is not necessary to enrich the entire domain with these R-functions.

Thus, the R-functions provide an effective step to build enrichment functions which help to impose boundary conditions on the domain boundary or internal interfaces on the domain without having a mesh in conformity with these boundaries.

3.4.2.2. *Neumann-type boundary conditions*

As for the application of Dirichlet boundary conditions, the application of boundary conditions of the Neumann type (typically of forces) requires particular processing if the domain boundary is not compatible with the quadtree description.

In the case of homogeneous boundary conditions, the integration of the weak form associated with the problem is immediate. It is only necessary to integrate this weak form on the portion of the cells in the field. For these cells, a weak form can be written as

$$\int_{\Omega} H(\boldsymbol{x}) \nabla^s \delta \boldsymbol{v} : \mathbf{C} : \nabla^s \boldsymbol{u} \mathrm{d}\Omega = \int_{\Omega} H(\boldsymbol{x}) \delta \boldsymbol{v} \cdot \boldsymbol{b} \mathrm{d}\Omega \quad \forall \delta \boldsymbol{v} \in \mathcal{V}, \qquad [3.30]$$

where $H(\boldsymbol{x})$ is the Heaviside function which takes the value 1 inside the domain and 0 outside the domain.

For cells cutting the boundary with non-homogeneous boundary conditions, the term relating to the external forces must be integrated separately and added to the second member (the column of the external generalized efforts $\boldsymbol{f}^{\text{ext}}$) of the final matricial system. It typically concerns the term $\int_{\Gamma_t} \bar{\boldsymbol{t}} \cdot \boldsymbol{u}^* \mathrm{d}\Gamma$ of equation [2.27], which takes part in the second member of the matricial equation [2.31]. For this, initially, the boundary is locally approached by a segment – dashed line in Figure 3.32 – and the integration is then carried out on this segment owing, for example, to a two-point Gaussian diagram.

3.4.2.3. *Partition of unity method*

Essentially, the partition of unity method uses a collection of sub-domains $\{\Omega_i\}$, which cover the domain completely. In the context of a quadtree approach, the Ω sub-domains are defined by the low level of the cells. On each one of these sub-domains, a function V_i representing the solution (in a certain sense) is built. Finally, let $\{\phi_i\}$ be a partition of unity respecting this recovery. In this case, the partition of unity is defined

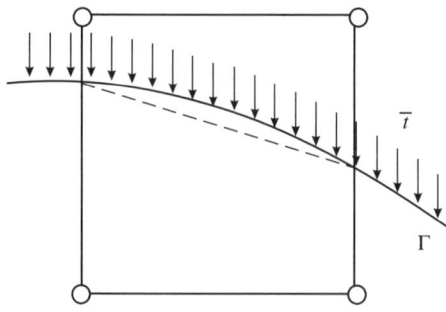

Figure 3.32. *Cell crossed by a boundary with natural boundary conditions*

by the restriction of the natural neighbor shape functions to the considered cell. As discussed earlier, the sum of the natural neighbor coordinates of a point with respect to its neighbor is 1.

The main contribution of the partition of unity is that the approximation space introduced, V, defined by

$$V = \sum_i \phi_i V_i \tag{3.31}$$

inherits the properties of the approximation spaces V_i and of the regularity of the partition of unity ϕ_i, and that while having introduced an inherited decomposition of the quadtree structure.

Chapter 4

Applications in the Mechanics of Structures and Processes

In this chapter, we examine the applications of the natural element method (NEM) for various constitutive equations, encountered both in structural mechanics and in the simulation of processes.

The behaviors considered are based on elasticity, thermo-viscoplasticity, and thermo-visco-elastoplasticity in large transformations. Eulerian and Lagrangian formulations (or updated Lagrangian) are also discussed in this chapter. Finally, the estimation of errors and the associated strategies of refinement are discussed.

4.1. Two- and three-dimensional elasticity

To validate the effectiveness of the NEM approach in 3D, we consider the problem of a circular tube subjected to an internal pressure. The geometry of the problem is defined in Figure 4.1. The analytical solution for this problem is available in the majority of works discussing elasticity, such as [TIM 72]. This solution is expressed in the polar coordinate system (ρ, θ, z):

$$\sigma_\rho = \frac{R_i^2 p}{R_e^2 - R_i^2}\left(1 - \frac{R_e^2}{\rho^2}\right), \tag{4.1a}$$

$$\sigma_\theta = \frac{R_i^2 p}{R_e^2 - R_i^2}\left(1 + \frac{R_e^2}{\rho^2}\right), \tag{4.1b}$$

$$\sigma_z = \nu\left(\sigma_\rho + \sigma_\theta\right), \tag{4.1c}$$

$$\varepsilon_\rho = \frac{1}{E}\left(\sigma_\rho - \nu\sigma_\theta - \nu\sigma_z\right), \qquad\qquad\qquad [4.2a]$$

$$\varepsilon_\theta = \frac{1}{E}\left(\sigma_\theta - \nu\sigma_\rho - \nu\sigma_z\right), \qquad\qquad\qquad [4.2b]$$

$$\varepsilon_z = 0, \qquad\qquad\qquad\qquad\qquad\qquad\qquad [4.2c]$$

$$u_\rho = \frac{R_i^2 p\rho}{E\left(R_e^2 - R_i^2\right)}\left[1 - \nu + \frac{R_e^2}{\rho^2}\left(1 + \nu\right)\right],$$

$$u_\theta = 0, \qquad\qquad\qquad\qquad\qquad\qquad\qquad [4.3]$$

$$u_z = 0,$$

where R_i and R_e, respectively, are the internal and external radii of the tube, and p is the internal pressure applied. The problem can be treated as a problem of plane strain. For this purpose we fixed the displacement of the bases of the tube to guarantee the state of plane strain. Moreover, as the symmetry of the problem is given, only the gray area of Figure 4.1 is analyzed (Figure 4.3). The three nodal distributions of Figure 4.2 are discussed, which contain 166, 241, and 1,708 nodes, respectively. The elastic properties of the materials are $E = 1.0$ and $\nu = 0.25$.

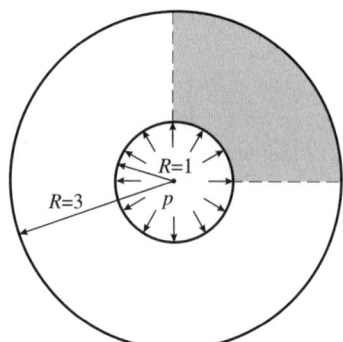

Figure 4.1. *Geometry of the problem of a circular tube subjected to an internal pressure*

The discretization is carried out by considering a Sibson interpolation of displacements (defined in 3D), and then the application of the Galerkin technique, where the numerical integration of the variational formulation is made with a quadrature with four points for each tetrahedron. The errors (norm in energy) are calculated with the same numerical quadrature. The results are grouped in Table 4.1.

The results relating to convergence are shown in Figure 4.4. The 3D results are equivalent to those obtained in [SUK 98a] by considering a state of plane strain (2D).

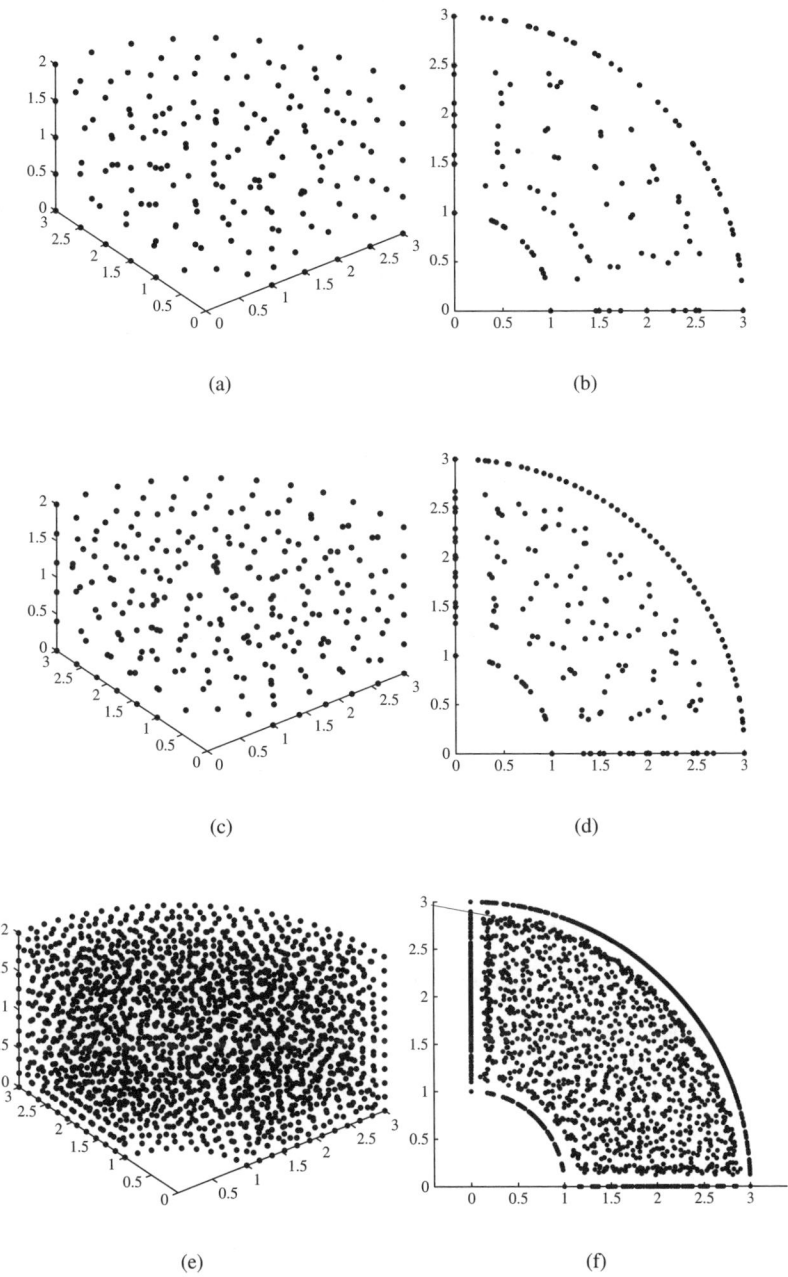

Figure 4.2. *Nodal distributions that are discussed*

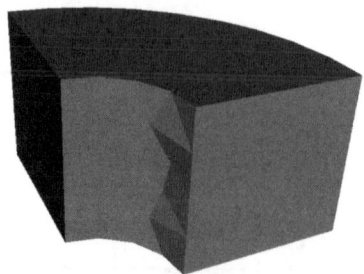

Figure 4.3. *Form associated with the cloud with 166 nodes*

Method	Number of nodes	$\|e\|_{L_2}$	$\|e\|_E$
α-NEM	166	4.74×10^{-3}	1.90×10^{-1}
FEM	166	6.36×10^{-3}	3.41×10^{-1}
α-NEM	241	2.88×10^{-3}	1.86×10^{-1}
FEM	241	3.51×10^{-3}	3.10×10^{-1}
α-NEM	2,076	1.12×10^{-3}	7.05×10^{-2}
FEM	2,076	9.56×10^{-4}	1.37×10^{-1}

Table 4.1. *Results associated with the problem of a circular tube subjected to an internal pressure*

The fundamental characteristics of the NEM in 3D seem to be a generalization of those obtained in 2D [CUE 02].

4.2. Indicators and estimators of error: adaptivity

4.2.1. *Meshless methods and adaptation*

One of the advantages of the meshless method is to tolerate great distortions of the starting node cloud without introducing significant errors into the solution. However, it is obvious that in all cases, a suitable nodal density is necessary to describe the strong gradients that are able to intervene with the solution of the problem. This is why an adaptation of the nodal density is necessary to calculate the numerical solution of the problems controlled by the partial differential equations.

In the context of the finite element method (FEM), these adaptations are included in the operation of remeshing. Remeshing is necessary, for example, when the geometry of the elements becomes too deformed after great deformation of the domain. Moreover, it is sometimes necessary to improve the quality of the

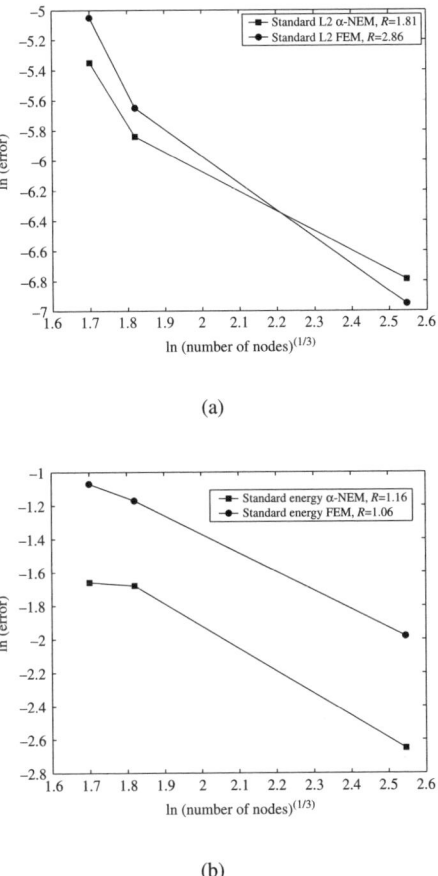

(a)

(b)

Figure 4.4. *Convergence of the problem of a circular tube subjected to an internal pressure*

interpolation in the vicinity of a strong gradient in the solution to add new nodes or to reposition them. This is not a simple task because the mesh associated with the new nodal distribution should not contain distorted elements. The geometrical quality of the elements depends on a constraint that largely complicates the refinement procedures, especially in 3D cases.

However, in meshless methods, there is no geometrical stress imposed on the relative position of the nodes. Thus, the introduction, elimination, or repositioning of nodes is a common task. In this manner, nodes can be added without geometrical checks in the areas where the quality of the solution must be improved (identified by a suitable error indicator). Once the new nodes have been placed in the domain and if the problem uses historical variables then the initialization of these variables within the framework of a nodal integration (all the variables are then defined in the nodes) can be realized in a common way by using the *meshless* interpolation scheme. This interesting aspect simplifies the refinement procedures significantly.

To determine the areas where refinement is necessary, error estimators or indicators are required. If the error estimate is well drawn within the framework of the FEM [LAD 01, BAB 79, ZIE 88], the emergence of publications on these subjects in the context of the meshless methods becomes more recent. In the context of *Reproducing Kernel Particle Method (RKPM)*, Belytschko *et al.* [BEL 98b] and Liu *et al.* [LIU 97] have proposed error estimates based on the residues of the equilibrium equation. You *et al.* [YOU 03] proposed an original approach using the RKPM approximation as a low-pass filter, and the corresponding high-pass filter is used to locate the areas with strong gradients, similar to the multi-resolution analysis and the wavelet analysis. In the context of the *Element Free Galerkin (EFG)* method, the error estimate was greatly studied using indicators based on the procedures of rebuilding (recovery methods) [ZIE 88]. As the shape functions possess great continuity, the stress reconstruction procedure does not require projection as in the FEM that uses C^0 shape functions. Through this, Chung and Belytschko [CHU 98] have introduced local and total error estimates into the EFG method. The essential factor of this error indicator is to use the difference between the values obtained by stress reconstruction and those obtained directly by the EFG solution. A thorough study was carried out by Lee and Zhou [LEE 03b, LEE 03a], and also by Lu and Chen [LU 02] for the RKPM. However, some problems occur because the quality of the efficiency indicator depends on the number of nodes contained in the moving least squares support function used for calculation and projection [CHU 98]. Moreover, the difficulties associated with the adaptation of the size of the shape functions support appear during the refinement process, as discussed in [LEE 03a]. These difficulties are related to the fact that the size of the support of the moving least squares shape functions must be regulated so that it covers a sufficient number of nodes. This is without, however, containing too large a number to preserve the quality. Thereafter, we show that the use of natural element shape functions makes it possible to avoid these difficulties.

4.2.2. *Methodology of adaptive refinement for linear elasticity in statics*

As this section is limited to linear elasticity, we propose an adaptive refinement procedure usable in the context of the constrained natural element method (CNEM) method including two error indicators a posteriori, both based on the technique of stress reconstruction based on the Zienkiewicz type and Zhu [ZIE 88], and based on the refinement technique based on Voronoi cells to allow refining in irregular node clouds.

4.2.2.1. *Formulation in static linear elasticity*

Let us consider the 2D problem of linear elasticity (the extension to 3D is immediate) expressed by the equilibrium equation:

$$\nabla \cdot \boldsymbol{\sigma} + \mathbf{b} = 0 \quad \text{in } \Omega, \qquad [4.4]$$

where $\Omega \subset \mathbb{R}^2$ is the material domain, ∇ the operator gradient, $\boldsymbol{\sigma}$ the stress tensor of the Cauchy stresses, and $\mathbf{b}$ is a term of body force.

The constitutive equation is given by

$$\boldsymbol{\sigma} = \mathbf{C}\boldsymbol{\epsilon}, \qquad [4.5]$$

where $\boldsymbol{\sigma}$ and $\boldsymbol{\epsilon}$ are the shape columns of the components of the stress tensor and the linearized strain tensor (symmetrical part), respectively. These are written according to the conventional notations:

$$\boldsymbol{\sigma} = \left\{ \begin{array}{c} \sigma_{11} \\ \sigma_{22} \\ \sigma_{12} \end{array} \right\} \quad \boldsymbol{\epsilon} = \left\{ \begin{array}{c} \epsilon_{11} \\ \epsilon_{22} \\ 2\epsilon_{12} \end{array} \right\},$$

with $\mathbf{C}$ as the elastic tensor in its matrix form. The essential and natural boundary conditions are given by

$$\begin{aligned} \mathbf{u} &= \bar{\mathbf{u}} \quad \text{on } \Gamma_u, \\ \boldsymbol{\sigma} \cdot \mathbf{n} &= \bar{\mathbf{t}} \quad \text{on } \Gamma_t, \end{aligned} \qquad [4.6]$$

where $\Gamma = \Gamma_u \cup \Gamma_t$ is the boundary of the domain Ω, $\mathbf{n}$ is the unit outward normal Γ, $\bar{\mathbf{u}}$ and $\bar{\mathbf{t}}$ are the imposed displacements and tractions, respectively.

The variational formulation (principle of virtual work) associated with the elastostatic problem is given by:

To find $\mathbf{u} \in H^1(\Omega)$ *kinematically acceptable* $(\mathbf{u} = \bar{\mathbf{u}}$ *on* $\Gamma_u)$ *such as*

$$\int_{\Omega} \boldsymbol{\sigma} \cdot \boldsymbol{\epsilon}^* \, d\Omega = \int_{\Omega} \mathbf{v}^* \cdot \mathbf{b} \, d\Omega + \int_{\Gamma_t} \mathbf{v}^* \cdot \bar{\mathbf{t}} \, d\Gamma, \quad \forall \mathbf{v}^* \in H_0^1(\Omega), \qquad [4.7]$$

where $H^1(\Omega)$ and $H_0^1(\Omega)$ are the usual Sobolev functional spaces.

By substituting approximation and test functions (both approximated by natural neighbors shape functions) in the preceding equation and by using the fact that field $\mathbf{v}^*$ is arbitrary, we obtain the system of linear equations after numerical integration:

$$\mathbf{K}\mathbf{d} = \mathbf{f}^{\text{ext}}, \qquad [4.8]$$

where **d** is the column containing nodal displacements, with the matrix **K** given by:

$$\mathbf{K} = \int_{\Omega} \mathbf{B}^t \mathbf{C} \mathbf{B} \, \mathrm{d}\Omega.$$ [4.9]

We derive

$$\mathbf{f}^{\text{ext}} = \int_{\Gamma_t} \mathbf{N}^t \bar{\mathbf{t}} \, \mathrm{d}\Gamma + \int_{\Omega} \mathbf{N}^t \mathbf{b} \, \mathrm{d}\Omega.$$ [4.10]

N is the matrix containing the shape functions:

$$\mathbf{N} = \begin{Bmatrix} \phi_1 & 0 & \phi_2 & 0 & \cdots & \phi_N & 0 \\ 0 & \phi_1 & 0 & \phi_2 & \cdots & 0 & \phi_N \end{Bmatrix}.$$

B is the matrix containing the derivative of CNEM shape functions:

$$\mathbf{B} = \begin{Bmatrix} \phi_{1,x} & 0 & \phi_{2,x} & 0 & \cdots & \phi_{N,x} & 0 \\ 0 & \phi_{1,y} & 0 & \phi_{2,y} & \cdots & 0 & \phi_{N,y} \\ \phi_{1,y} & \phi_{1,x} & \phi_{2,y} & \phi_{2,x} & \cdots & \phi_{N,y} & \phi_{N,x} \end{Bmatrix}.$$ [4.11]

The matrix **C** of elastic linear isotropic behavior is given in plane strain by

$$\mathbf{C} = \frac{E}{(1-2\nu)(1+\nu)} \begin{Bmatrix} 1-\nu & \nu & 0 \\ \nu & 1-\nu & 0 \\ 0 & 0 & \dfrac{1-2\nu}{2} \end{Bmatrix},$$

and in the case of plane stress by

$$\mathbf{C} = \frac{E}{(1-\nu^2)} \begin{Bmatrix} 1 & \nu & 0 \\ \nu & 1 & 0 \\ 0 & 0 & \dfrac{1-\nu}{2} \end{Bmatrix}.$$

The use of a stabilized conforming nodal integration (SCNI) leading to a strain tensor given by

$$\bar{\epsilon}^h(\mathbf{x}_i) = \frac{1}{A_i} \int_{\Omega_i} \epsilon^h(\mathbf{x}) \, \mathrm{d}\Omega = \frac{1}{A_i} \int_{\Omega_i} \begin{Bmatrix} \dfrac{\partial u^h(\mathbf{x})}{\partial x} \\ \dfrac{\partial v^h(\mathbf{x})}{\partial y} \\ \dfrac{\partial u^h(\mathbf{x})}{\partial y} + \dfrac{\partial v^h(\mathbf{x})}{\partial x} \end{Bmatrix} \mathrm{d}\Omega.$$ [4.12]

By applying the theorem of divergence, we obtain equation [4.13]

$$\tilde{\epsilon}^h\left(\mathbf{x}_i\right) = \frac{1}{A_i} \int_{\Gamma_i} \left\{ \begin{array}{c} u^h(\mathbf{x})n_x \\ v^h(\mathbf{x})n_y \\ u^h(\mathbf{x})n_y + v^h(\mathbf{x})n_x \end{array} \right\} d\Gamma, \qquad [4.13]$$

where Ω_i is the representative domain associated with the node n_i and Γ_i and A_i are the boundaries and the surface (volume in 3D) of the representative domain, respectively.

By introducing the NEM approximation introduced into equation [4.13] we obtain

$$\tilde{\epsilon}^h\left(\mathbf{x}_i\right) = \tilde{\mathbf{B}}_i\mathbf{d}, \qquad [4.14]$$

which is given explicitly by

$$\tilde{\epsilon}^h\left(\mathbf{x}_i\right) = \left\{ \begin{array}{ccccc} \tilde{b}_{x1}\left(\mathbf{x}_i\right) & 0 & \cdots & \tilde{b}_{xN}\left(\mathbf{x}_i\right) & 0 \\ 0 & \tilde{b}_{y1}\left(\mathbf{x}_i\right) & \cdots & 0 & \tilde{b}_{yN}\left(\mathbf{x}_i\right) \\ \tilde{b}_{y1}\left(\mathbf{x}_i\right) & \tilde{b}_{x1}\left(\mathbf{x}_i\right) & \cdots & \tilde{b}_{yN}\left(\mathbf{x}_i\right) & \tilde{b}_{xN}\left(\mathbf{x}_i\right) \end{array} \right\} \left\{ \begin{array}{c} u_1 \\ v_1 \\ u_2 \\ v_2 \\ \vdots \\ u_N \\ v_N \end{array} \right\} = \tilde{\mathbf{B}}_i\mathbf{d}, \quad [4.15]$$

where $\mathbf{d}$ is the column of all the degrees of the freedom interpolation and $\tilde{b}_{xj}(\mathbf{x}_i)$ and $\tilde{b}_{yj}(\mathbf{x}_i)$ are defined by

$$\tilde{b}_{xj}\left(\mathbf{x}_i\right) = \frac{1}{A_i} \int_{\Gamma_i} \phi_j(\mathbf{x})n_x(\mathbf{x})\, d\Gamma, \qquad [4.16]$$

$$\tilde{b}_{yj}\left(\mathbf{x}_i\right) = \frac{1}{A_i} \int_{\Gamma_i} \phi_j(\mathbf{x})n_y(\mathbf{x})\, d\Gamma. \qquad [4.17]$$

A great number of terms in the matrix $\tilde{\mathbf{B}}_i$ are zero. This is due to the fact that the CNEM shape functions have a compact support. Chen *et al.* [CHE 01] proposed to use the intersection of Voronoi cells as representative domains Ω_i with the domain. By introducing the CNEM approximation, stabilized strain and nodal integration, the global stiffness matrix is obtained by assembling the contribution of each node n_k:

$$\mathbf{K} = \sum_i \tilde{\mathbf{K}}_i = \sum_i A_i \tilde{\mathbf{B}}_i^t \mathbf{C} \tilde{\mathbf{B}}_i. \qquad [4.18]$$

It is possible to associate an "average" stress with each cell obtained from derivatives that are stabilized by

$$\tilde{\sigma}_i^h = \mathbf{C} \tilde{\mathbf{B}}_i\mathbf{d}. \qquad [4.19]$$

4.2.2.2. *The first indicator based on the rebuilding of discontinuous stress fields*

In the FEM, the field reconstruction procedure proposed by Zienkiewicz and Zhu [ZIE 88] consists of comparing the stress field provided by the calculation result (discontinuous with the passage of the elements) with a more continuous reconstructed field (obtained by a projection of moving least squares type), supposed to provide a solution of better quality than the FE solution.

In the NEM, the solution provided by the calculated result is, however, very regular, even in the passing of one Voronoi cell to another. However, by using an average gradient per cell, nodal integration (section 3.3.2) degrades this continuity and provides a discontinuous stress field to the passage of the cells. By analogy with the Zienkiewicz and Zhu's method, we use the discontinuous fields obtained by nodal integration like the field solution. The reconstructed field (of reference) is given by a NEM-type interpolation of the stress nodal values. The procedure is detailed in the following section.

Let $\mathbf{u}^h$ be the nodal displacements obtained from the linear system resolution [4.8] resulting from discretization NEM of the variational formulation expressed by equation [4.7]. Through the application of SCNI as described in section 3.3.2, a constant stress field in each cell (discontinuous on the edges of Voronoi cells) can be calculated from $\tilde{\varepsilon}_i^h$ (equation [4.15]). Thus, the stress field can be obtained in each Voronoi cell as

$$\tilde{\boldsymbol{\sigma}}_i^h = \mathbf{C}\tilde{\varepsilon}_i^h = \mathbf{C}\tilde{\mathbf{B}}_i\mathbf{u}^h, \qquad [4.20]$$

with $\tilde{\mathbf{B}}_i$ as defined in equation [4.15]. Moreover, we define in every point $\mathbf{x} \in \Omega$ the reconstructed stress field $\hat{\boldsymbol{\sigma}}(\mathbf{x})$ by interpolating the nodal values $\tilde{\boldsymbol{\sigma}}_i^h$ resulting from the calculations obtained by the NEM shape functions. This reconstructed field has the same continuity as the natural interpolation

$$\hat{\boldsymbol{\sigma}}(\mathbf{x}) = \sum_{i=1}^{V} \phi_i^C(\mathbf{x})\tilde{\boldsymbol{\sigma}}_i^h, \qquad [4.21]$$

where V is the number of visible natural neighbors from points $\mathbf{x}$ and ϕ_i^C are CNEM shape functions and built on the basis of the constrained Voronoi diagram (section 2.3.2). Differentiation between these two fields can be used as a local error indicator which is represented by $NN1$. The local error indicator is used by means of a standard in energy that takes the following form:

$$NN1 : \tilde{e}_i^h = \left[\int_{\Omega_i} \frac{1}{2}\left(\tilde{\boldsymbol{\sigma}}_i^h - \hat{\boldsymbol{\sigma}}\right)^T \mathbf{C}^{-1}\left(\tilde{\boldsymbol{\sigma}}_i^h - \hat{\boldsymbol{\sigma}}\right) d\Omega \right]^{1/2} = \left\| \tilde{\boldsymbol{\sigma}}^h - \hat{\boldsymbol{\sigma}} \right\|_i, \qquad [4.22]$$

where Ω_i is the constrained Voronoi cell associated with the node n_i.

4.2.2.3. *The second indicator based on the rebuilding of continuous fields of strain*

Again, we use the methodology from the preceding section, but by substituting the discontinuous stress field with a continuous stress field obtained in any point starting from the derivative of the shape functions. This field is defined in equation [4.23]:

$$\boldsymbol{\sigma}^h(\mathbf{x}) = \mathbf{C}\mathbf{B}^h(\mathbf{x})\mathbf{u}^h, \qquad\qquad [4.23]$$

where $\mathbf{B}^h(\mathbf{x})$ is a matrix containing the derivative of the shape functions calculated as point $\mathbf{x}$. It should be noted, however, that $\boldsymbol{\sigma}^h(\mathbf{x})$ is not defined for $\mathbf{x} = \mathbf{x}_i$.

The second error indicator can now be obtained like differentiation between the fields defined in equations [4.21] and [4.23]. We call this indicator $NN2$.

$$NN2 : e_i^h = \left[\int_{\Omega_i} \frac{1}{2} (\boldsymbol{\sigma}^h - \widehat{\boldsymbol{\sigma}})^T \mathbf{C}^{-1} (\boldsymbol{\sigma}^h - \widehat{\boldsymbol{\sigma}}) d\Omega \right]^{1/2} = \left\| \boldsymbol{\sigma}^h - \widehat{\boldsymbol{\sigma}} \right\|_i. \qquad [4.24]$$

Let it be noted, however, that the indicator $NN2$ can simply be implemented within the framework of linear problems.

4.2.2.3.1. Effectivity indexes

To numerically evaluate the integrals defined in equations [4.22] and [4.24], the Voronoi cells are triangulated (Figure 4.5), and a Gaussian integration is applied in each triangle.

By using Voronoi cells as a partition of the domain, the total error can be calculated as

$$\left(\tilde{e}_\Omega^h \right)^2 = \sum_{i=1}^{N} \left(\tilde{e}_i^h \right)^2 \qquad\qquad [4.25]$$

and

$$\left(e_\Omega^h \right)^2 = \sum_{i=1}^{N} \left(e_i^h \right)^2, \qquad\qquad [4.26]$$

where N is the number of nodes in the domain. The local contributions to the total error associated with indicators $NN1$ and $NN2$ can also be defined as

$$\tilde{\eta}_i^h = \frac{\tilde{e}_i^h}{\|\widehat{\sigma}\|_\Omega}, \qquad\qquad [4.27]$$

$$\eta_i^h = \frac{e_i^h}{\|\widehat{\sigma}\|_\Omega}. \qquad\qquad [4.28]$$

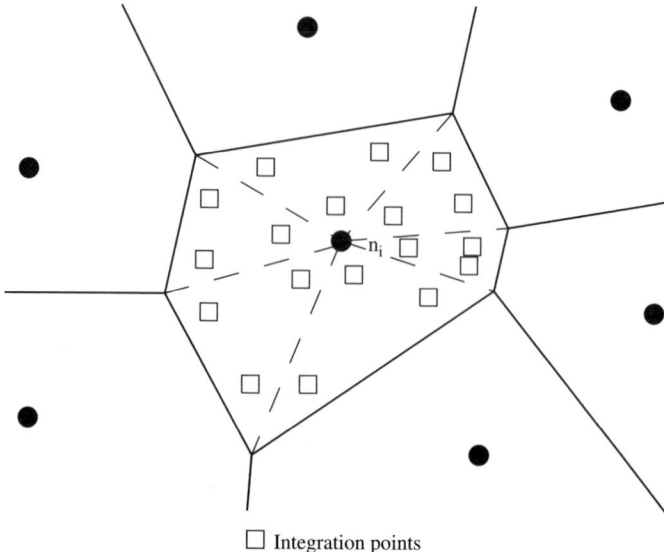

☐ Integration points

Figure 4.5. *Integration of the error in each cell*

The associated efficiency markers θ are defined as the relationship between the estimated error and the exact error:

$$\theta_{NN1} = \tilde{\theta}^h = \frac{\tilde{e}^h_\Omega}{e^{ex}_\Omega} = \frac{\left\|\tilde{\sigma}^h - \hat{\sigma}\right\|_\Omega}{\left\|\tilde{\sigma}^h - \sigma^{ex}\right\|_\Omega}, \qquad [4.29]$$

$$\theta_{NN2} = \theta^h = \frac{e^h_\Omega}{e^{ex}_\Omega} = \frac{\left\|\sigma^h - \hat{\sigma}\right\|_\Omega}{\left\|\sigma^h - \sigma^{ex}\right\|_\Omega}. \qquad [4.30]$$

The efficiency of a good indicator should be close to 1 (or at least in an asymptotic way when the nodal density increases). As in Zienkiewicz *et al.* [ZIE 99], we can establish the following bound for indicator $NN1$:

$$\left\|\tilde{\sigma}^h - \hat{\sigma}\right\|_\Omega = \left\|(\tilde{\sigma}^h - \sigma^{ex}) - (\hat{\sigma} - \sigma^{ex})\right\|_\Omega, \qquad [4.31]$$

hence

$$\left\|\tilde{\sigma}^h - \sigma^{ex}\right\|_\Omega - \left\|\hat{\sigma} - \sigma^{ex}\right\|_\Omega \leq \left\|\tilde{\sigma}^h - \hat{\sigma}\right\|_\Omega \leq \left\|\tilde{\sigma}^h - \sigma^{ex}\right\|_\Omega + \left\|\hat{\sigma} - \sigma^{ex}\right\|_\Omega, \quad [4.32]$$

and by dividing each side with $\left\|\tilde{\sigma}^h - \sigma^{ex}\right\|_\Omega$ we can obtain equation [4.33]

$$1 - \frac{\left\|\hat{\sigma} - \sigma^{ex}\right\|_\Omega}{\left\|\tilde{\sigma}^h - \sigma^{ex}\right\|_\Omega} \leq \theta_{NN1} \leq 1 + \frac{\left\|\hat{\sigma} - \sigma^{ex}\right\|_\Omega}{\left\|\tilde{\sigma}^h - \sigma^{ex}\right\|_\Omega}. \qquad [4.33]$$

We can write a similar relationship for the associated efficiency index $NN2$. From equation [4.33] we can conclude that if the convergence solution $\hat{\sigma}$ has a higher rate than that of the approximated solution $\tilde{\sigma}^h$, the estimate of the error must be asymptotically exact.

4.2.2.4. Refinement strategy based on Voronoi cells

To improve the quality of the approximation without detriment to the calculation time, that is, to adapt the nodal density to the solution of the problem, the knowledge of the local error contribution is particularly important. For this, we can define a total permissible error $\bar{\eta}$ and a permissible error in each cell $\bar{\eta}_i$ satisfying the relationship

$$\bar{\eta}^2 = \sum_{i=1}^{N} \left(\bar{\eta}_i\right)^2.$$ [4.34]

If we homogenize the errors in the domain, the permissible error $\bar{\eta}_i = \bar{\eta}_*$, $\forall i$ in each cell can be given by:

$$\bar{\eta}_* = \frac{\bar{\eta}}{\sqrt{N}}.$$ [4.35]

Thereafter, we use the efficiency index that is close to 1 to qualify $\bar{\eta}$ as the error for reasons of simplicity. By following Zienckiewicz et al. [ZIE 99], we propose to calculate the characteristic length of the new cell compared with the cell having generated it by assuming a p-order convergence $(O((h^p)))$. We have then, for example, for the indicator $NN2$:

$$\frac{h_i^{\text{new}}}{h_i^{\text{old}}} = \left[\frac{\bar{\eta}_*}{\eta_i}\right]^{\frac{1}{p}} = \left[\frac{\bar{\eta}\|\hat{\sigma}(\mathbf{x})\|_\Omega}{\sqrt{N}e_i^h}\right]^{\frac{1}{p}}.$$ [4.36]

From our experience, the rate of convergence in the approximations using the natural shape function elements are very close to those obtained by the use of linear finite elements (due to the linear consistency of the Sibson interpolation). Thus, p should be the rate of convergence by using an energy norm.

To decide whether a new cell should be divided or not, let us suppose that the characteristic length of the new cells generated by the aforementioned method are twice as small (Figure 4.6), for reasons of simplicity,

$$\frac{h^{\text{new}}}{h^{\text{old}}} \leq \frac{1}{2}.$$ [4.37]

We can thus determine whether a cell should be subdivided when there is

$$\left[\frac{\overline{\eta}\|\widehat{\sigma}(\mathbf{x})\|_\Omega}{\sqrt{N}e_i^h}\right]^{\frac{1}{p}} \leq \frac{1}{2}.$$ [4.38]

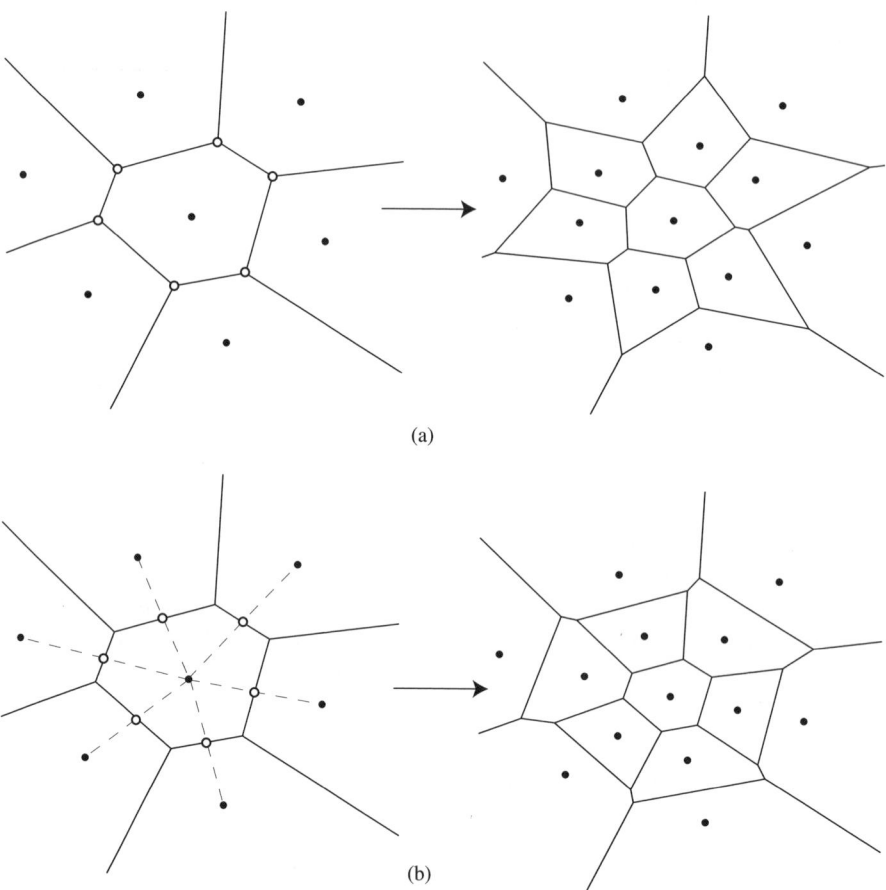

Figure 4.6. *Refinement strategies based on Voronoi cells – (o): extra nodes*

The strategy represented in Figure 4.6(a) consists of adding nodes on the vertices of the Voronoi cell when the error in a cell exceeds the limit described in equation [4.38]. This strategy was used by several authors, such as [YOU 03] and [LU 02], which indeed made it possible to refine irregular node clouds. The strategy illustrated in Figure 4.6(b) consists of adding (for a cell whose error exceeds the limiting value) the nodes between the natural node n_i and the neighbors n_j. We recommend the use of strategy in Figure 4.6(b) because in the case of quasi-regular grids, the orthocenters on which the new nodes will be positioned can be very close to each other, even stacking up to make regular grids. Strategy in Figure 4.6(b) is equivalent to a subdivision of the elements in the FEM.

It is noted that in the NEM, with all entities associated with the Voronoi diagram already constructed for the needs of interpolation, no additional construction is introduced for the needs of interpolation, which is contrary to the moving least squares method. Moreover, we cannot find the adjustment problem for the shape function's support size. This is because it is automatically defined by its natural neighbors which naturally vary with density and local anisotropy.

4.3. Metal extrusion

The numerical simulation of the extrusion methods, as with other forming processes, is of great importance to avoid trial correction procedures during the early stages. The first simulations of the extrusion methods date from 1960s (see [ZIE 84] and its references). During the last decades, the FEM was largely used in the simulation of extrusion procedures. The term "flow formulation" was proposed by Zienkiewicz *et al.* [ZIE 78b, ZIE 74, ZIE 78a].

Despite the existence of an elastic return at the die exit, the aforementioned procedure is often ignored to simplify the resolution of the thermomechanical model. This model greatly resembles those resulting from non-Newtonian viscous flow models; hence, the name is given as flow formulation [ZIE 74].

The equations which control the deformation of the metal is thus written in terms of velocity field as opposed to displacement field often used in structural mechanics. The Cauchy stress tensor can be expressed by

$$\boldsymbol{\sigma} = \boldsymbol{D}(\boldsymbol{d}, T) \cdot \boldsymbol{d}, \qquad [4.39]$$

where $\boldsymbol{d}$ represents the strain rate tensor (symmetric part of the tensor gradient rates) and T the temperature.

Recently, Bonet [BON 99] proposed a formulation, for this type of non-Newtonian flow, using the displacement field. The author termed this alternative as "formulation of incremental flow." The ability to take incompressibility into account is the main advantage of such a formulation.

The majority of researchers supposed (see [ZIE 84] for example) the stationarity of the flow, a hypothesis that makes use of a fixed mesh and also an Eulerian simulation possible. The Eulerian approach can also be used in transient mode, on a condition that certain corrections will be introduced (often introduced in an iterative way) [LOF 01].

However, the use of an Eulerian formulation, despite the advantages in terms of managing the mesh, presents other disadvantages. Let us take, for example, the heat equation in a Lagrangian context

$$\rho c_p \frac{dT}{dt} + \nabla(k\nabla T) + \dot{r} = 0, \qquad [4.40]$$

where k represents thermal conductivity (presumed here as isotropic), $\dot{r}$ the source term that represents plastic dissipation during the working process, ρ the voluminal mass and c_p the specific heat of material. Thus, if we choose an Eulerian description instead of a Lagrangian one, the heat equation can be written as

$$\rho c_p \left(\frac{\partial T}{\partial t} + \boldsymbol{v}\nabla T \right) + \nabla(k\nabla T) + \dot{r} = 0. \qquad [4.41]$$

It is well known that in this case the convective term which results from Euler's writings on the material derivative requires a suitable stabilization during its discretization. The difficulty related to convective stabilization is absent from the Lagrangian formulation frameworks.

However, the main difficulty linked with Lagrangian descriptions comes from consecutive mesh distortions to large strains which emerge during the materials forming. If today we have high-performance 2D re-meshing tools at our disposal, this is not the case for 3D tools. Many researchers have preferred to employ *Arbitrary Lagrangian Eulerian* (ALE) procedures to reduce this problem. In this description, the mesh is transported with a velocity different from the material velocity. The heat equation [4.40], which results from it, takes the form:

$$\rho c_p \left(\frac{\partial T}{\partial t} + (\boldsymbol{v} - \overline{\boldsymbol{v}})\nabla T \right) + \nabla(k\nabla T) + \dot{r} = 0, \qquad [4.42]$$

where $\overline{\boldsymbol{v}}$ represents the velocity of the mesh. Evidently, if $\overline{\boldsymbol{v}} = \boldsymbol{v}$ we obtain the Lagrangian description. On the contrary, if $\overline{\boldsymbol{v}} = \boldsymbol{0}$ we obtain the Eulerian description.

The main objective of the ALE formulation is to minimize the distortion of the mesh and thus the re-meshing stages. Some examples with regard to the application of the ALE techniques in the extrusion simulation processes were presented by Huétink and other contributors [ATZ 95, MOO 95, LOF 99, LAN 00]. Other information can also be found in [LOF 02, LOF 01, LOF 00]. The main difficulty lies in the velocity determination of mesh $\overline{\boldsymbol{v}}$. However, even if the term of convection is less problematic than in purely Eulerian descriptions, the convective term still requires stabilization during its discretization.

4.3.1. *Viscoplastic model for the extrusion of aluminum*

In this section, we disregard the elastic return. The selected model is therefore purely viscoplastic. The variables of the model are velocity and pressure.

The deviatoric stress tensor can be written as

$$s = 2\mu d, \qquad [4.43]$$

by having the classic relationship

$$\sigma = s - pI, \qquad [4.44]$$

where $p = -tr(\sigma)/3$ and I is the indentity tensor. Generally, the coefficient μ depends on the intensity of strain rate (making the model nonlinear) and the temperature.

To derive the expression from the parameter μ, the strain rate tensors are calculated from a viscoplastic potential. While following the model of Perzyna [PER 66], we obtain

$$d^{vp} = \dot{\gamma}\frac{\partial Y(\sigma, q)}{\partial \sigma}, \qquad [4.45]$$

where Y is the viscoplastic potential which coincides, in general, with the criterion of plasticity, and $\dot{\gamma}$ is the scalar

$$\dot{\gamma} = \frac{\langle g(Y(\sigma, q))\rangle}{\eta} \text{ with } \langle x \rangle = \frac{x + |x|}{2}, \qquad [4.46]$$

$\langle g \rangle$ is a monotonic function that takes zero value only if $Y(\sigma, q) \leq 0$, η is a positive parameter often called viscosity, and q represents hardening. In the following, we remove the annotation by exposing ^{vp} that we previously used to refer to viscoplastic behavior.

For metals, and particularly for aluminum, the criterion for the most usual flow comes from Mises:

$$Y(\sigma, q) = \overline{\sigma}(s) - \sigma_y(\overline{d}, T), \qquad [4.47]$$

where $\overline{\sigma} = \sqrt{\frac{3}{2}s : s} = \sqrt{3J_2}$ represents the effective stress and σ_y the uniaxial stress flow. $\overline{d}$ is the only internal variable that represents the rate of effective strain:

$$\overline{d} = \sqrt{\frac{2}{3}d : d}. \qquad [4.48]$$

The most commonly used law to describe the aluminum yield stress comes from Sellars–Tegart [SEL 72]:

$$\sigma_y(\overline{d}) = S_m \text{arcsinh}\left[\left[\left(\frac{\overline{d}}{A}\right)e^{\frac{Q}{RT}}\right]^{\frac{1}{m}}\right]. \qquad [4.49]$$

with S_m, m, and A being the material parameters. It should be noted that, as one can expect the yield stress decreases when the temperature (thermal softening) increases.

If we consider this model, a stress rate of zero leads to a yield stress of zero. This does not correspond to the general behavior considered for metals, particularly aluminum. To free itself from this deficiency, an up-to-date method consists of amending equation [4.49] to include an initial strain rate $\bar{d}_0$ leading to Sellars–Tegart modified law:

$$\sigma_y(\bar{d}) = S_m \text{arcsinh}\left[\left[\left(\frac{\bar{d}_1}{A}\right)e^{\frac{Q}{RT}}\right]^{\frac{1}{m}}\right] \text{ with } \bar{d}_1 = \max\left\{\bar{d}, \bar{d}_0\right\}. \qquad [4.50]$$

If we combine the general shape of the tensor strain rates for equation [4.45] with equation [4.47], we obtain

$$d = \dot{\gamma}\frac{3s}{2\bar{\sigma}}. \qquad [4.51]$$

If we now combine equation [4.48] with the expression of real stress $\bar{\sigma}$, we obtain that $\dot{\gamma}$ is the precise real strain rate:

$$\bar{d} = \frac{\dot{\gamma}}{\bar{\sigma}}\sqrt{\frac{3}{2}s : s} = \dot{\gamma}. \qquad [4.52]$$

However, by considering Perzyna's model used in equations [4.45] and [4.46] and considering that $g(f) = f$, we can obtain the expression which connects the equivalent stress and equivalent strain rate:

$$\bar{d} = \dot{\gamma} = \frac{\langle\bar{\sigma} - \sigma_y\rangle}{\eta} \implies \bar{\sigma} = \eta\bar{d} + \sigma_y \text{ if } \bar{\sigma} \geq \sigma_y, \qquad [4.53]$$

which, introduced into equation [4.51], and taking equation [4.52] into account, leads to the viscoplastic constitutive equation:

$$s = 2\frac{\eta\bar{d} + \sigma_y(\bar{d})}{3\bar{d}}d. \qquad [4.54]$$

According to the return value, it should be noted that the yield surface takes place at a different rate

$$s = \frac{2\sigma_y}{3\bar{d}}d. \qquad [4.55]$$

Finally, taking into account the incompressibility of the plastic flow, the constitutive equation is written as equation [4.56]:

$$\sigma = 2\mu d - pI, \text{ with } \mu = \frac{\sigma_y}{3\bar{d}}. \qquad [4.56]$$

Of course this simple model has important limitations, with the absence of elasticity being one of the most important. Thus, the elastic return cannot be predicted. In spite of everything, this type of modeling has made it possible to obtain wholly acceptable simulations of different methods [ZIE 74, ZIE 78b, LOF 02].

This model type has also been fully used in the polymer forming framework, leading to a constitutive equation known as pseudoplastics

$$\mu = \mu_0 \overline{\dot{d}}^{n-1}, \qquad [4.57]$$

where n is the sensitivity coefficient and μ_0 the consistence coefficient. For other examples of such applications, see [SUK 05].

4.3.2. 3D simulation of the extrusion of a cross-shaped profile

In this section, we discuss the geometry that can be found in [ZHO 03]. The geometry of the die is represented in Figures 4.7 and 4.8. The diameter of the cylinder to be extruded measures 50 mm and its 20 mm length makes it possible to reach the stationary mode. The numerical model contains 1,313 nodes distributed in one area of the domain benefitting from symmetrical conditions on the $X = 0$ and $X = Y$ planes. The initial nodal distribution is shown in Figure 4.9.

The nodes located on the upper plane are forced to move with the speed of $2 \, \text{mms}^{-1}$ so as to obtain an exit velocity of approximately $1 \, \text{m min}^{-1}$. We took a sliding condition between the material and the mould. The initial temperature is fixed at 723K. From the thermal point of view the mould is considered adiabatic. In the same way, we neglected convective cooling on the extruded surface. All the thermomechanical parameters characterizing aluminum are listed in Table 4.2. Sellars–Tegart's ideal rigid plastic that we previously discussed was used in simulations.

The simulation was carried out for more than 42 time steps (each one of $0.025 \, s$). The evolution of the equivalent strain rate is represented in Figure 4.10. Figures 4.11 and 4.12 represent the evolution of pressure. Despite having used a Sibson–Thiessen approximation that does not verify the LBB stability condition, once again the pressure distribution appears satisfactory.

The third variable of the model, such as the temperature, is represented in Figures 4.13 and 4.14. As can be expected, the highest temperatures are present when the strain velocity and the heat generation are at a maximum.

Finally, it can be noted that the yield stress evolution represented in Figure 4.15 is at its lowest when the temperatures are at a maximum. This is in agreement with the Sellars–Tegart model that was discussed earlier.

Figure 4.7. *Geometry of the die*

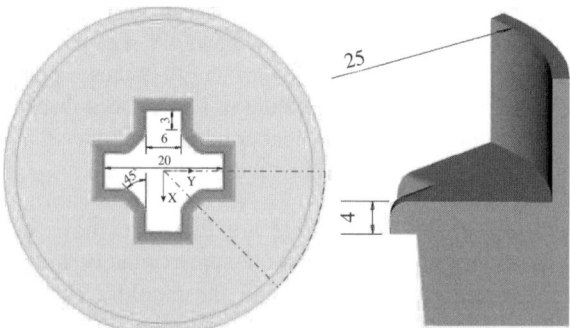

Figure 4.8. *Dimensions of the die (mm) – the contour of the simulated area is indicated by dotted lines*

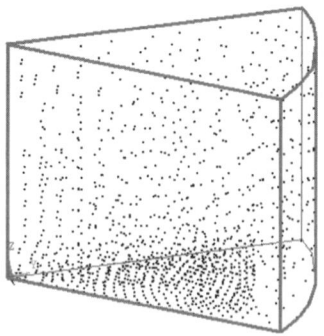

Figure 4.9. *Initial distribution of the nodes*

Parameter	Units	Value
$\overline{d}_0$	(s^{-1})	0.005
S_m	(MPa)	25
m		5.4
A	(s^{-1})	6×10^9
Q	$(\frac{J}{mol})$	1.4×10^5
R	$(\frac{J}{mol\,K})$	8.314

Table 4.2. *Material parameters of aluminum alloy AA6063 (values taken from [LOF 02])*

It should be noted that the calculated temperatures are lower than those obtained in the extrusion method. This difference is likely to be due to the limits of the sliding conditions, which have been discussed. To assess the effect of the condition of contact on the calculated temperatures, we carry out the simulation again by imposing a perfect adhesion condition (instead of a perfect slip condition) to the tool material interfaces. The impact on the profile velocity is shown in Figure 4.16. The existence of a "dead" area comprising a recirculating flow is highlighted. The same effect can be observed in the temperature field as shown in Figure 4.17.

4.4. Friction stir welding

"Friction stir welding" (FSW) is a method of welding especially well adapted to assembling pieces of poor weldability. The method consists of welding the parts in contact via a pin in rotation, which softens the material. This is achieved by heat created by the friction, which is generated between the material and the rotating pin. Rotation allows a well-blended mix of the materials on the two sides of the initial interface. The rotating pin is thus activated by rotary motion as well as by translation motion all along the interface which needs to be welded. The method is represented schematically in Figure 4.18. The material is then moved from the front of the pin toward the back. Once solidified, it ensures a close joint of the two adjoining parts.

From a numerical viewpoint, this method hides many difficulties. These are linked to large strains, large strain velocity due to high velocity of the rotating pin and to the long simulation time taken due to the weak velocity of the pin advancement. Finally, these difficulties are linked to the way in which the blending of the materials is handled when put in contact with both sides of the interface.

To our knowledge, the number of works to date devoted to the simulation of such a method is rather reduced. In [CHE 06], the use of a Lagrangian description was

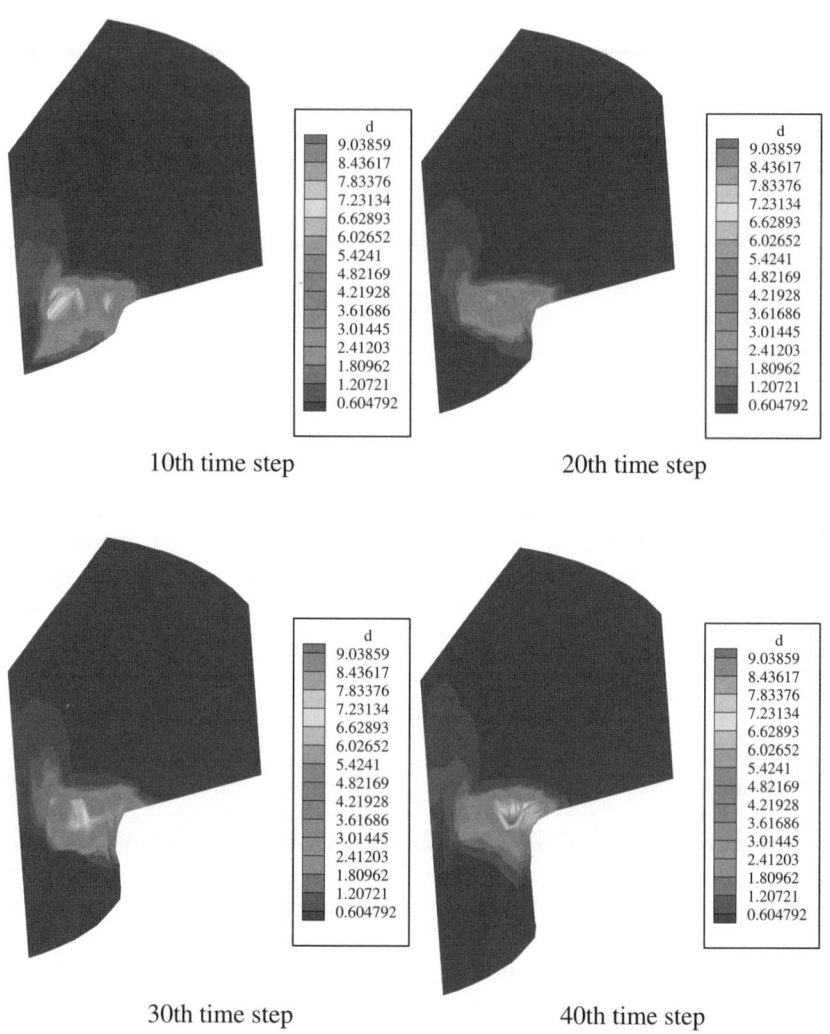

Figure 4.10. *Equivalent strain rate* (s^{-1})

proposed with intensive meshing and similar approaches were employed in [BUF 06a] and [BUF 06b]. However, the reader has already been forewarned with regard to intensive 3D meshing. To eliminate problems related to meshing, a formulation based on the use of the NEM is implemented again within the framework of an updated Lagrangian description. This formulation makes it possible to work with the same cloud of nodes during the entire simulation, while avoiding nodal repositioning.

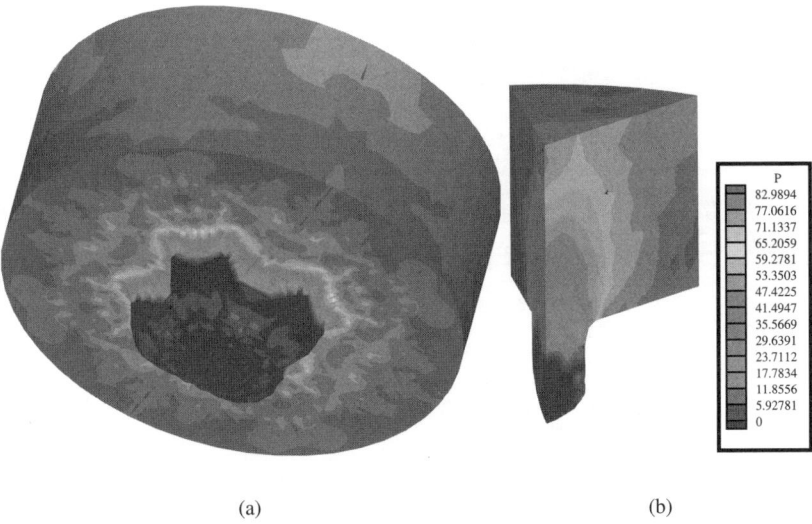

Figure 4.11. *Pressure (N/mm²) in the 20th time step*

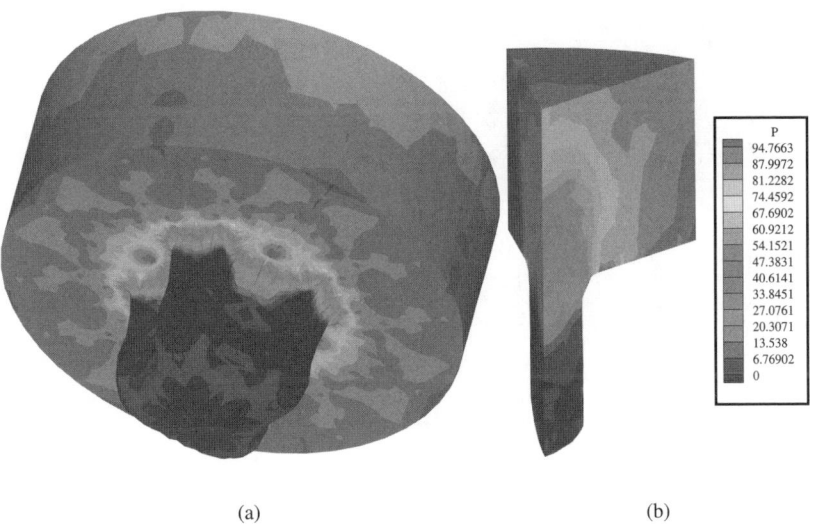

Figure 4.12. *Pressure (N/mm²) in the 40th time step*

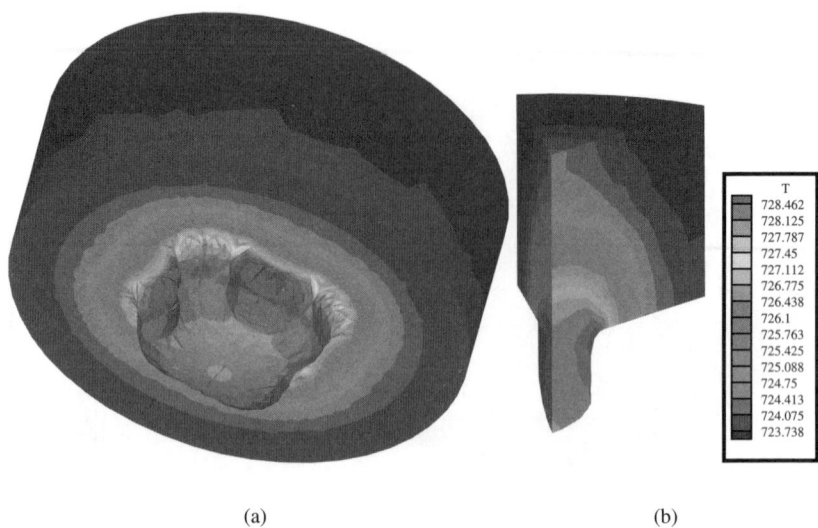

(a)   (b)

Figure 4.13. *Distribution of temperatures (K) in the 20th time step*

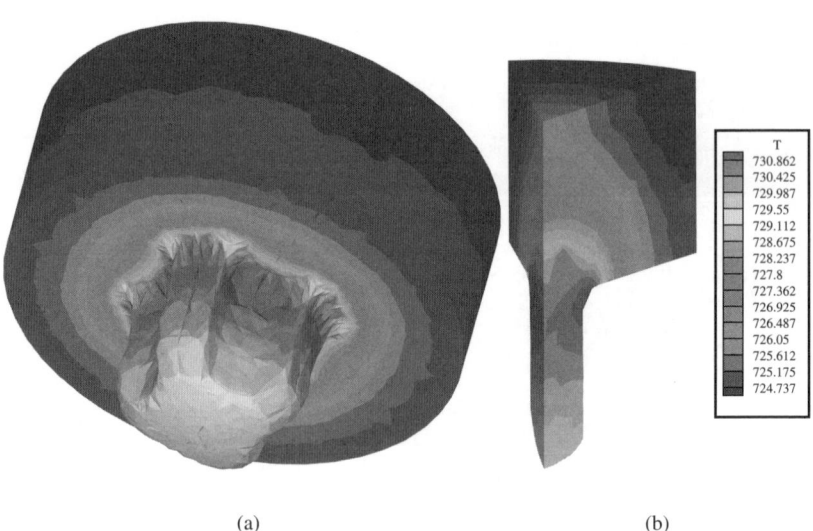

(a)   (b)

Figure 4.14. *Temperature distribution (K) in the 40th time step*

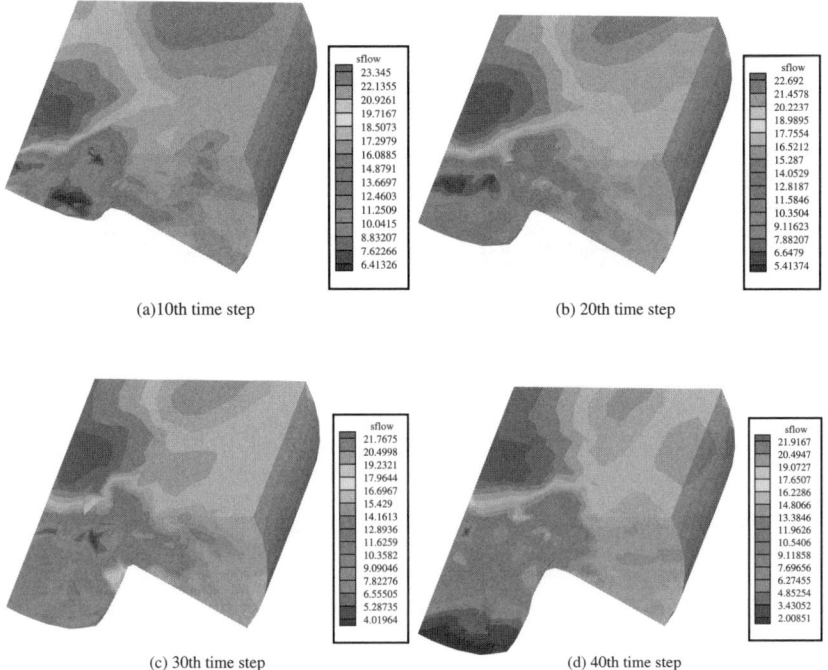

(a)10th time step

(b) 20th time step

(c) 30th time step

(d) 40th time step

Figure 4.15. *Plastic flow stress (N/mm^2)*

Therefore, the projection stage of the internal variables between the different meshes at the beginning of numerical diffusion becomes important.

4.4.1. *Constitutive equation*

The important plastic strain present in stir friction welding justifies neglecting the elasticity. Therefore, simulations are greatly simplified at the price of not being able to give an account of possible elastic returns. The constitutive equation thus behaves similar to the simulation of extrusion described in the preceding section (equation [4.56]).

In this section, we consider the yield stress expressed by

$$\sigma_y = \mathrm{K}\, T^A \, \overline{d}^B \, \overline{\varepsilon}^C, \qquad\qquad [4.58]$$

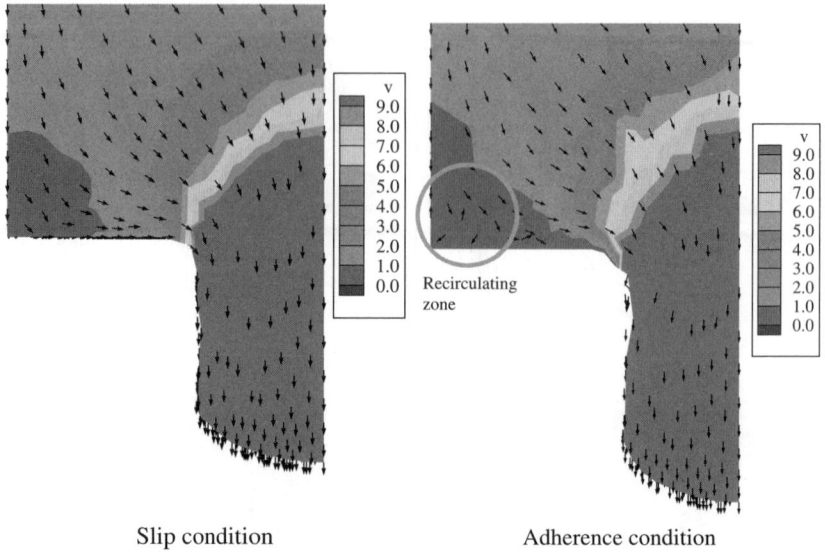

Slip condition   Adherence condition

Figure 4.16. *Comparison between velocity profiles with slip condition and adhesion condition on the tool–material interfaces (mm/s)*

where $K = 2.69 \times 10^{10}$, $A = -3.3155$, $B = 0.1324$, and $C = 0.0192$. In the following simulations, we do not consider the influence of strain into account so that $C = 0$. This law has previously been employed in other works, such as [BUF 06a].

By following the method described in the section on the simulation of extrusion, we obtain the following constitutive equation:

$$\boldsymbol{\sigma} = 2\mu\mathbf{d} - p\mathbf{I}, \text{ with } \mu = \frac{\sigma_y}{3\bar{\bar{d}}}. \qquad [4.59]$$

4.4.1.1. *Mechanical model*

In the following, we consider momentum equations, when the effects of inertia are disregarded:

$$\boldsymbol{\nabla} \cdot \boldsymbol{\sigma} = \mathbf{0}. \qquad [4.60]$$

The incompressibility of the plastic strain is written in turn as

$$\boldsymbol{\nabla} \cdot \mathbf{v} = 0, \qquad [4.61]$$

where $\mathbf{v}$ represents the velocity field.

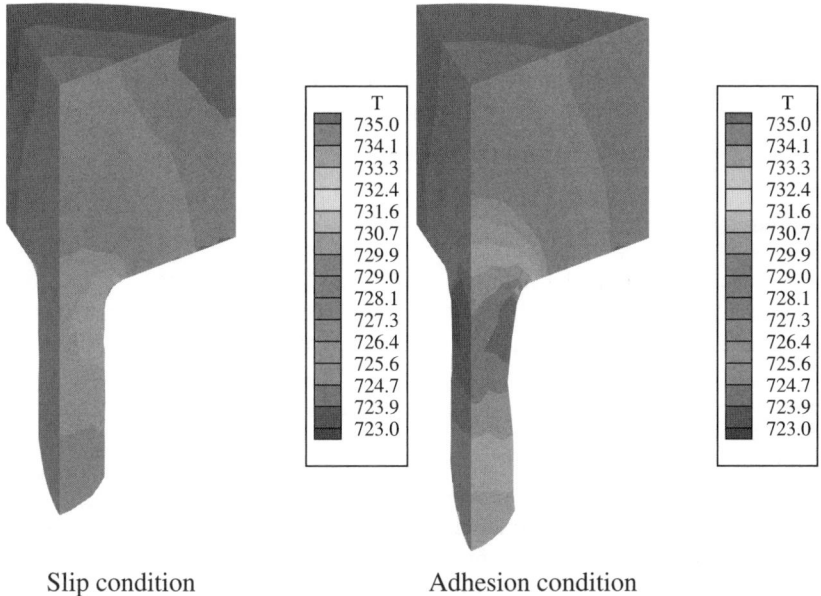

Slip condition   Adhesion condition

Figure 4.17. *Comparison between temperatures (K) with slip and adhesion conditions on the tool–material interfaces*

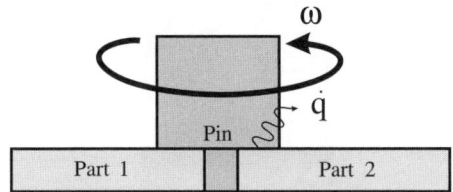

Figure 4.18. *Schematic representation of the stir friction welding method*

The thermal effects are described by

$$\nabla(k\nabla T) + \dot{r} - \left(\rho c_p \dot{T}\right) = 0, \qquad [4.62]$$

where k is the thermal conductivity, $\dot{r}$ the source term, ρ the specific density, and c_p the specific heat. The source term due to plastic strain takes the form

$$\dot{r} = \beta\boldsymbol{\sigma} : \mathbf{d}, \qquad [4.63]$$

where β represents the fractional mechanical energy transformed into heat, which is often in the order of $\beta \approx 0.9$ [ZHO 03].

The coupling is carried out by ways of a semi-implicit strategy, with a fixed-point algorithm for the resolution of the nonlinear mechanical model. This strategy was used and described in detail in [ALF 06a].

4.4.2. *Numerical results*

We consider a 2D model here, in plane stress, concerning the welding of two plates. The model comprises 240 nodes. The Voronoi diagram is recalculated at every time step and after updating the nodal positions with material velocities resulting from the mechanical problem resolution. The Voronoi diagram reconstruction is extremely quick if we compare its cost against the cost of the mechanical problem resolution.

The pin rotates $1,000$ tr/min, moving 1 mm/s. Even if a friction coefficient of approximately 0.5 is generally selected (0.46 in [BUF 06a]), here we consider a perfect adhesion between the rotating pin and material comprising the plates to be welded. The time step is fixed here at 10^{-4} seconds so as not to compromise the stability.

Figures 4.19–4.22 show the change in temperature at various time steps. Figures 4.23–4.26 represent the equivalent strain rate. In each case, simulations are in line, at least on the qualitative front, with those obtained by [BUF 06a] with the FEM.

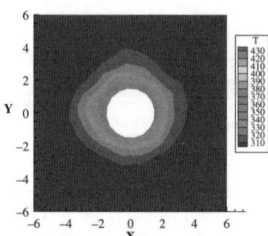

Figure 4.19. *Field of temperature in the 10th time step*

Without a doubt, the main interest in the technique that we have used is the possibility of following each node during its history. We can therefore access the mix of materials to be welded. Indeed, a label is associated with each node, indicating the material it represents at the beginning of the simulation. This label remains unchanged throughout the simulation because the formulation is Lagrangian. This is also because in our approach it is not necessary to move the nodes with respect to the matter for remeshing the geometrical quality standards for the elements. For each time step, we

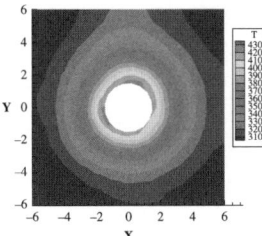

Figure 4.20. *Field of temperature in the 50th time step*

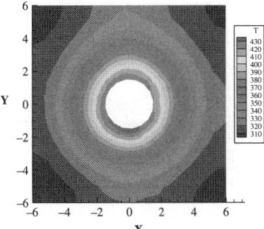

Figure 4.21. *Field of temperature in the 70th time step*

can therefore follow the labels on the different nodes and therefore access the map of materials at each moment.

In Figure 4.27, the initial configuration is represented where the nodes appear in light or dark gray according to their respective plates. The simulation requires approximately 4,000 time steps. The nodal configuration after these 4,000 time steps is represented in Figure 4.28.

The technique of following nodes throughout the simulation eliminates the need for projections and makes it possible to retain the identity of each node. This makes it possible to follow the mix of materials and avoids any type of numerical diffusion related to the field transfers between the old and the new meshes. Figure 4.29 gives the Delaunay triangulation to the 4,000th time step. In spite of the existence of extreme deformed triangles that cannot be used in the context of finite elements, these distortions by no means compromise the quality of the results obtained with the NEM.

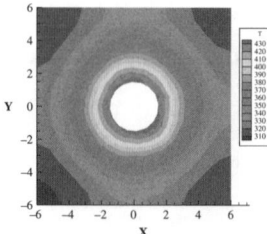

Figure 4.22. *Field of temperature in the 90th time step*

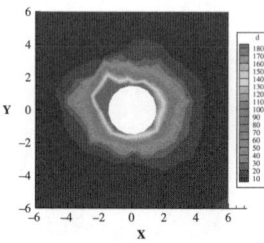

Figure 4.23. *Equivalent strain rate to the 10th time step*

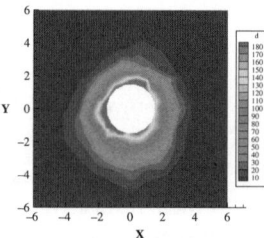

Figure 4.24. *Equivalent strain rate to the 50th time step*

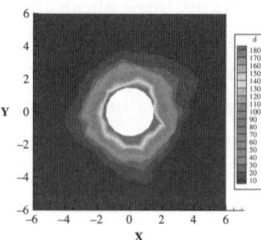

Figure 4.25. *Equivalent strain rate to the 70th time step*

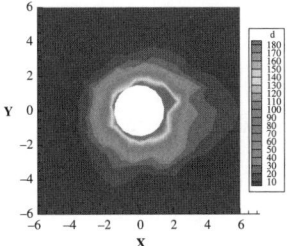

Figure 4.26. *Equivalent strain rate in the 90th time step*

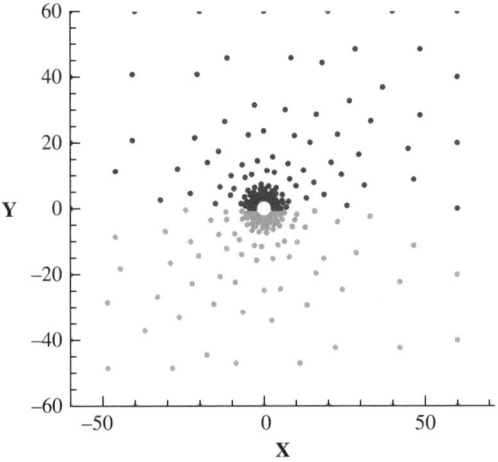

Figure 4.27. *Initial setup*

4.5. Models and numerical treatment of the phase transition: foundry and treatment of surfaces

4.5.1. *Introducing motion discontinuity*

In this section, we demonstrate that the CNEM makes it possible to introduce motion discontinuities in a simple way into a fixed node cloud in whichever way is better, by simple insertion or removal of a new inner boundary. As no criterion on the relative position of the nodes is necessary, a simple updating of the stressed Voronoi diagram is carried out after the insertion of the interface. The appearance of flattened Delaunay triangles does not degrade, which is shown through the examples of phase transition problems.

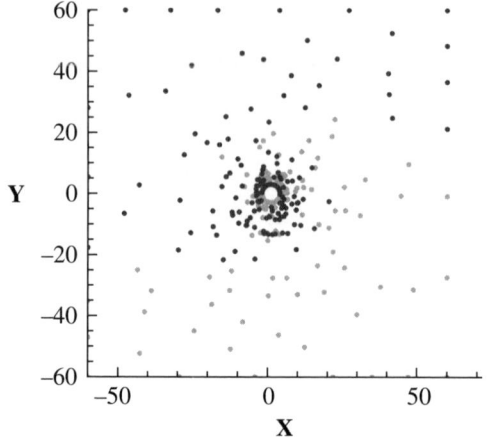

Figure 4.28. *Nodal distribution after 4,000 time steps*

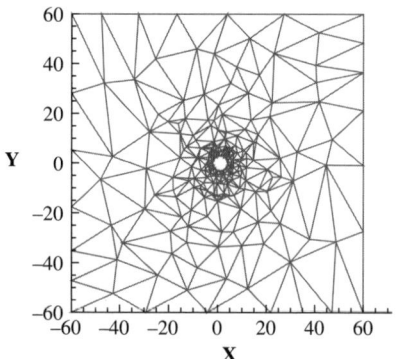

Figure 4.29. *Delaunay triangulation in the 4,000th time step*

4.5.1.1. *Formulation of the thermal problem with phase transition*

Let $\Omega \subset \mathbb{R}^2$ be a bounded domain and T a temperature field. We consider Γ_1 as the boundary of the domain where the temperature is imposed $T(\mathbf{x} \in \Gamma_1, t) = \overline{T}(\mathbf{x}, t)$, and Γ_2 as the boundary of the domain where the thermal flux $\overline{q}$ is imposed. The thermal model is defined in a time interval $[0, t_{\max}]$. The initial temperature $T(\mathbf{x}, t = 0) = T_0$, where T_0 is calculated above the melting point T_m. When $t = 0$, a part Γ_1 of the boundary of the domain is suddenly subjected to a temperature $T_1 < T_m$. A moving solidification front $\Gamma_I(t)$ is then generated, with a changing position through time, which divides the domain Ω into two ($\Omega_1(t)$ containing the solid phase at time t and $\Omega_2(t)$ containing the liquid phase), represented in Figure 4.30. For simplicity, hereafter we consider a model of isotropic and homogeneous conduction in each phase.

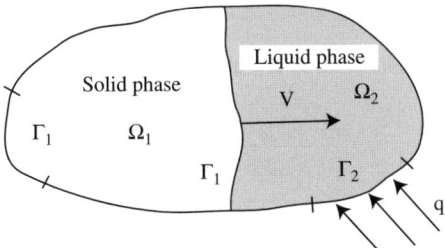

Figure 4.30. *Biphasic thermal problem*

The model of heat conduction is defined in each phase, neglecting volumetric source terms, by

$$\begin{cases} c_1 \dfrac{\partial T(\mathbf{x}, t)}{\partial t} = \nabla \cdot \left(k_1 \nabla T \right) & \text{in } \Omega_1(t) \\[2mm] c_2 \dfrac{\partial T(\mathbf{x}, t)}{\partial t} = \nabla \cdot \left(k_2 \nabla T \right) & \text{in } \Omega_2(t), \end{cases}$$
[4.64]

where c_1 and c_2 are the volumetric heat capacities in each phase, and k_1 and k_2 are the respective thermal conductivities. The initial and respective boundary conditions are

$$\begin{cases} T(\mathbf{x}, \, t = 0) = T_0 & \forall \mathbf{x} \in \Omega \\[1mm] T(\mathbf{x}, t) = \overline{T}(\mathbf{x}, t) & \forall \mathbf{x} \in \Gamma_1, \, \forall t \in \left[0, t_{\max} \right] \\[1mm] -k \nabla T(\mathbf{x}, t) \cdot \mathbf{n} = \bar{q}(\mathbf{x}, t) & \forall \mathbf{x} \in \Gamma_2, \, \forall t \in \left[0, t_{\max} \right]. \end{cases}$$
[4.65]

The evolution of the interface $\Gamma_I(t)$ is described by the Stefan condition:

$$\mathbf{V}\left(\mathbf{x} \in \Gamma_I(t) \right) = \frac{[[q]]}{L} \mathbf{n}_I(\mathbf{x})$$
[4.66]

where $\mathbf{V}$ is the velocity of the interface, L the latent volumetric heat of fusion, $\mathbf{n}_I(\mathbf{x})$ the normal unitary vector at the interface $\mathbf{x}$, turning from solid to liquid, and $[[q]]$ is the jump of the heat flux toward the interface $\Gamma_I(t)$, expressed by

$$[[q]] = \left(k_1 \nabla T \big|_{\Gamma_I^-(t)} - k_2 \nabla T \big|_{\Gamma_I^+(t)} \right) \mathbf{n}_I.$$
[4.67]

The additional stress imposed on interface $\Gamma_I(t)$ is

$$T(\mathbf{x}, t) = T_m; \quad \forall \mathbf{x} \in \Gamma_I(t),$$
[4.68]

where T_m is the melting temperature.

4.5.1.2. *CNEM discretization*

In this section, the temperatures are approximated by using CNEM interpolation scheme:

$$T^h(\mathbf{x}) = \sum_{i=1}^{V} \phi_i^C(\mathbf{x}) T_i, \qquad [4.69]$$

where V is the number of visible natural neighbors since point $\mathbf{x}$ and ϕ_i^C are the CNEM shape functions associated with the ith neighbor node visible from $\mathbf{x}$. The calculation of the CNEM shape functions is similar to that of the NEM shape functions, wherein the previously introduced constrained Voronoi diagram (CVD) has been used. We previously demonstrated that the use of the constrained Voronoi diagram does not affect the properties of NEM interpolation, allowing the extension of the shape function linearity on any portion of the domain boundary, whether convex or non-convex.

If we define for t two CSDs corresponding to $\Omega_1(t)$ and $\Omega_2(t)$, both constrained by interface $\Gamma_I(t)$, the interpolated temperature field is of C^1 continuity everywhere, except for the nodes and when the interface $\Gamma_I(t)$ appears, where the aforementioned is only C^0. Thus, this interpolation seems to be suitable to simulate the Stefan problem.

To illustrate this interpolation property, we consider the situation illustrated in Figure 4.31, where $\mathbf{x}$ moves from Ω_1 toward Ω_2. If $\mathbf{x}$ is in Ω_1, the interpolated field is constructed from equation [4.69] by using the visible natural neighbors from $\mathbf{x}$ (Γ_I is presumed opaque). If $\mathbf{x}$ is in Γ_I, according to the preceding discussion, the interpolated field is strictly linear because it depends solely on the two neighbors located on Γ_I. Finally, when $\mathbf{x}$ is in Ω_2, the interpolated field is defined by using the visible neighbors from $\mathbf{x}$ (Γ_I being opaque). The interpolation continuity is guaranteed, but a discontinuity appears in the derivative of the field, because an abrupt change of visible natural neighbors intervenes with passage of the interface. We can naturally reproduce the continuity of the field in the domain, as well as the expected discontinuity of the derivative when the interface appears. By supposing $\Gamma_2 \equiv \Gamma_I$, the weak formulation linked to equation [4.64] is:

Finding $T \in H^1(\Omega)$ verifying $T = \bar{T}$ on Γ_1 such as:

$$\int_\Omega c(\mathbf{x}) \frac{\partial T}{\partial t} \delta T \, d\Omega = -\int_\Omega k(\mathbf{x}) \nabla T \cdot \nabla \delta T \, d\Omega + \int_{\Gamma_I(t)} [\![q]\!] \delta T \, d\Gamma, \quad \forall \delta T \in H_0^1(\Omega),$$

$$[4.70]$$

where $c(\mathbf{x}) = c_i$ if $\mathbf{x} \in \Omega_i$, $k(\mathbf{x}) = k_i$ if $\mathbf{x} \in \Omega_i$. $H^1(\Omega)$ and $H_0^1(\Omega)$ are Sobolev's usual functional spaces. Substituting the approximation function and the test function

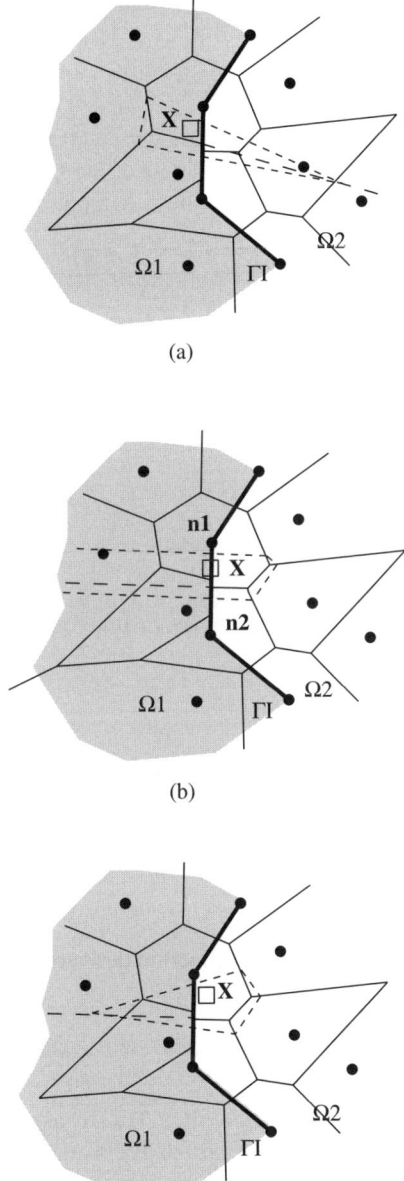

Figure 4.31. *Reproduction of discontinuous derivatives when the interface appears in the use of the constrained Voronoi diagram*

(both approximated by the diagram CNEM) in the preceding equation, and by using the fact that the field δT is arbitrary, the following system of equation is obtained:

$$\mathbf{C}\dot{\mathbf{T}} + \mathbf{K}\mathbf{T} = \mathbf{F}, \qquad\qquad [4.71]$$

where $\mathbf{T}$ is the column containing the unknown nodal temperatures. We consider the solution in the time interval $[0, t_{\max}]$, divided into time steps $[t^n, t^{n+1}]$ and the generalized trapezoidal time stepping algorithm characterized by the parameter α:

$$\frac{\partial T^{n+1}}{\partial t} = \frac{T^{n+1} - T^n - (1-\alpha)\Delta t \frac{\partial T^n}{\partial t}}{\alpha \Delta t}, \qquad\qquad [4.72]$$

which leads to

$$\left(\mathbf{C}^{n+1} + \alpha \Delta t \mathbf{K}^{n+1}\right)\mathbf{T}^{n+1} = \mathbf{F}^{n+1}\left(\mathbf{T}^n, |[q]|^{n+1}\right), \qquad\qquad [4.73]$$

with

$$\mathbf{C}^{n+1} = \int_{\Omega_1^{n+1}} \mathbf{N}^t c_1 \mathbf{N}\, d\Omega + \int_{\Omega_2^{n+1}} \mathbf{N}^t c_2 \mathbf{N}\, d\Omega, \qquad\qquad [4.74]$$

where $\Omega^{n+1} = \Omega_1^{n+1} \cup \Omega_2^{n+1}$;

$$\mathbf{K}^{n+1} = \int_{\Omega_1^{n+1}} \mathbf{B}^t k_1 \mathbf{B}\, d\Omega + \int_{\Omega_2^{n+1}} \mathbf{B}^t k_2 \mathbf{B}\, d\Omega \qquad\qquad [4.75]$$

and

$$\mathbf{F}^{n+1} = \mathbf{C}^{n+1}\mathbf{T}^n + (1-\alpha)\Delta t \int_{\Omega^{n+1}} \mathbf{N}^t c \frac{\partial T^n}{\partial t}\, d\Omega + \alpha \Delta t \int_{\Gamma_I^{n+1}} \mathbf{N}^t [[q]]^{n+1}\, d\Gamma, \qquad\qquad [4.76]$$

where $\mathbf{N}$ is the vector containing the nodal shape functions:

$$\mathbf{N} = \{\phi_1 \quad \phi_2 \quad \cdots \quad \phi_N,\}$$

and $\mathbf{B}$ is the matrix containing the shape function derivatives:

$$\mathbf{B} = \begin{Bmatrix} \phi_{1,x} & \phi_{2,x} & \cdots & \phi_{N,x} \\ \phi_{1,y} & \phi_{2,y} & \cdots & \phi_{N,y}. \end{Bmatrix}$$

The stabilized integration of Chen *et al.* in [CHE 01] as described in the preceding section is used for the numerical integration of $\mathbf{K}$. A diagonal matrix $\tilde{\mathbf{C}}$ is obtained naturally by using the surface areas of Voronoi constrained cells like nodal weights.

The incremental algorithm of resolution is defined in the following way:

Knowing $\mathbf{T}^n$ and $|[q]|^n$ at time t^n, the nonlinear problem associated with equation [4.73] is summarized to find $\mathbf{T}^{n+1}$ and $|[q]|^{n+1}$ such that equations [4.68] and [4.73] are satisfied. To this end, we proceed in the following way:

– Calculate the velocity of the interface $\mathbf{V}^n(\mathbf{x})$ by using equation [4.66] and to update the position of the interface at time t^{n+1} by using the forward Euler formula:

$$\mathbf{x}_J^{n+1} = \mathbf{x}_J^n + \Delta t \mathbf{V}^n\left(x_J^n\right),$$ [4.77]

where $\mathbf{x}_J$ are the nodes defining the interface.

– Locally update the constrained Voronoi diagram and the shape functions associated with the points of integration in the vicinity of the interface. Then, to solve equation [4.73] by using the Newton–Raphson method where the tangent matrix is evaluated numerically.

– Repeat as long as $t^{n+1} < t_{\max}$.

An alternative scheme using the Latin method [LAD 98] in the context of X-FEM was discussed by Merle and Dolbow [MER 02].

Example: unidirectional solidification of a semi-infinite solid.

In this section, we illustrate the potential of our technique to simulate a phase transition problem of the Stefan problem type. This problem is unidirectional, but here it is solved in 2D to highlight the characteristics of the method.

The Stefan problem models the one-dimensional solidification of a semi-infinite solid ($x \geq 0$). The initial temperature T_0 is assumed constant throughout the domain, and above the melting point T_m. When $t = 0$, the temperature on the left side $x = 0$ is abruptly subjected to value T_1 lower than the melting point, creating a solidification front which progresses from side $x = 0$ toward $x > 0$. The exact position of the front $x_f(t)$ is given by

$$x_f(t) = 2\lambda\sqrt{\beta_s t},$$ [4.78]

where $\beta_s = k_s/c_s$ is the thermal diffusivity of the solid phase, and the constant λ satisfies the relationship:

$$\frac{e^{-\lambda^2}}{\text{erf}(\lambda)} = \frac{k_l\sqrt{\eta}(T_0 - T_d)e^{-\eta\lambda^2}}{k_s(T_d - T_1)\text{erfc}(\lambda\sqrt{\eta})} + \frac{\lambda L\sqrt{\pi}}{c_s(T_d - T_1)},$$ [4.79]

with $\eta = \beta_s/\beta_l$ the ratio of the thermal diffusivities, where k_l represents the conductivity of the liquid phase. The field of temperature in the solid phase $0 \leq x \leq x_f(t)$ is then expressed by

$$T(x,t) = T_1 + \frac{T_m - T_1}{\text{erf}(\lambda)}\text{erf}\left(\frac{x}{2\sqrt{\beta_s t}}\right),$$ [4.80]

and in the liquid phase $x \geq x_f(t)$ by

$$T(x,t) = T_0 - \frac{T_0 - T_m}{\text{erfc}(\lambda\sqrt{\eta})}\text{erfc}\left(\frac{x}{2\sqrt{\beta_l t}}\right),$$ [4.81]

where erf and erfc are the mathematical functions error and complementary error, respectively.

In this study, we use the thermal properties provided in [LYN 81], which are listed in Table 4.3. T_1 and T_0 are initialized to -10 and $4.0\,^0C$, respectively ($\lambda = 0.3073$). We simulate the evolution of the domain of temperature in $\Omega = [0,1] \times [0,0.5]$ cm. To use the reference solution, given for an infinite domain, we set the temperature in $x = 1$ with the exact value given by equation [4.81].

Properties	Solid	Liquid
Volumetric capacity (cal. $^0C^{-1}$cm^{-3})	0.49	0.62
Thermal conductivity (cal. cm^{-1}s^{-1}.$^0C^{-1}$)	$9.6.10^{-3}$	$6.9.10^{-3}$
Melting point (^0C)		0.0
Volumetric latent heat of fusion (cal. cm^{-3})		19.2

Table 4.3. *Thermal properties of the material defined in [LYN 81]*

In the first test, we consider a domain Ω discretized by a uniform grid 20×10, with $\Delta t = 2\,s$. Figure 4.32 compares the calculated and the exact positions of the interface. We can see the good quality of the calculated solution, as shown in Figure 4.33 where the error in the position of the front is represented. Figure 4.34 shows the profile of temperature at different times. We can see that discontinuity in the temperature gradient is captured with precision. Good compliance with the exact solution is observed.

In the second test, we consider a domain Ω containing 200 nodes distributed randomly. The objective of this test is to evaluate the characteristics of the meshless technique in which no restriction on the relative position of the nodes is imposed. While placing the nodes in a random way, flattened Delaunay triangles can be formed (especially when the interface moves). Here we are seeking to determine whether the quality of the solution is quite independent of the position of nodes between them. Figure 4.35 represents the node cloud and the position of the interface, as well as the CVD. In spite of the totally random discretization, we can see that the interface remains rectilinear during its propagation through the domain. We note that the direction and the velocity of each point of the interface are not imposed, and depend only on the temperature field result as calculated previously. From Figures 4.33 and 4.36, we conclude that quality is effectively not altered by the irregularity of the nodal distribution. In Figure 4.37, temperature profiles along the line $y = 0.25$ are represented, showing a good compliance with the theoretical solution.

To study the convergence of the method, we proceeded to a series of calculations by gradually decreasing the distance between nodes. The results for the relative error on the position of the front when $t = 50$ are given in Figure 4.38, which shows the calculation convergence.

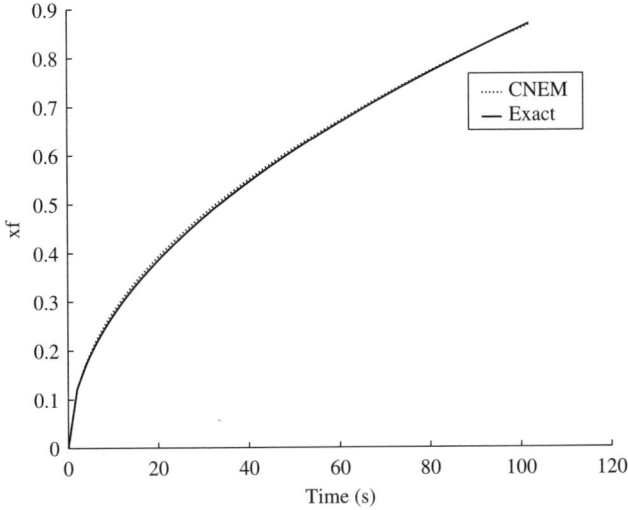

Figure 4.32. *Comparison between the theoretical front position and the CNEM solution using a* 20×10 *regular grid*

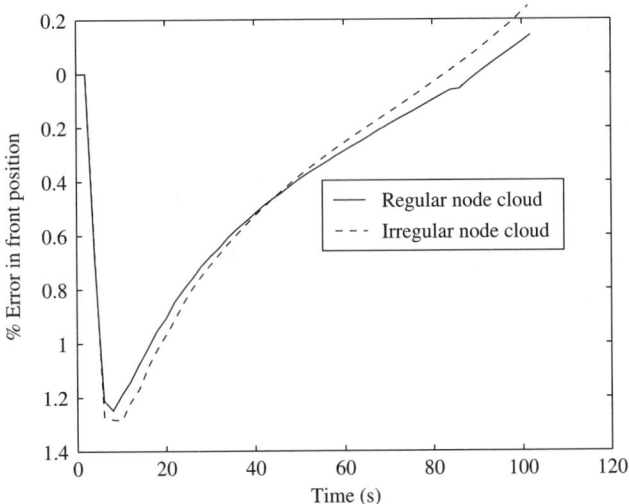

Figure 4.33. *Error in the calculated position of the front*

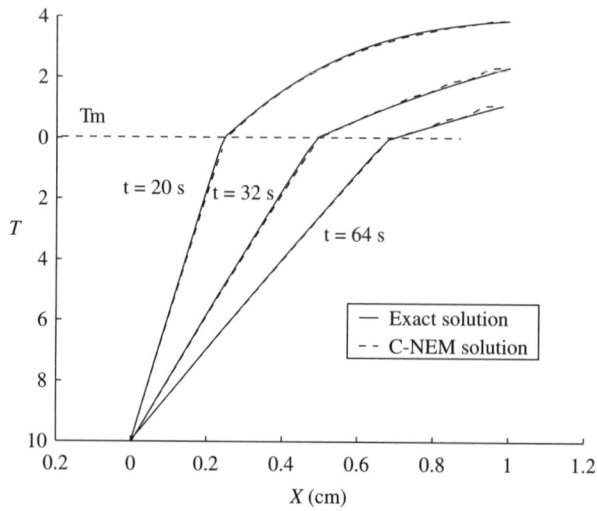

Figure 4.34. *Temperature profile along line $y = 0.25$ by using a regular grid*

Example: evolution of a radial face of solidification.

In this problem, a domain $\Omega = [-1, 1] \times [-1, 1]$ cm, initially entirely in the liquid state $(T_0 > T_m)$, is suddenly cooled in a point by a source of heat of localized negative value at the item $(0, 0)$, then creating an axisymmetric face of solidification separating the liquid phase and the solid phase whose radius increases with time. This problem was investigated previously by Ji *et al.* in [JI 02] in the context of X-FEM. The theoretical solution of this problem in an infinite domain of $\mathbb{R}^2$ can be found in [CAR 59]. The radius of the front of solidification is given by

$$R_f = 2\lambda\sqrt{\beta_s t}.$$ [4.82]

The temperature in the solid area $r < R_f$ is given by

$$T(r, t) = T_{\rm d} + \frac{Q}{4\pi k_s}\left[Ei\left(-\frac{r^2}{4\beta_s t}\right) - Ei\left(-\lambda^2\right)\right],$$ [4.83]

in the liquid area $r > R_f$ by

$$T(r, t) = T_0 - \frac{T_0 - T_{\rm d}}{Ei(\lambda^2\eta)}Ei\left(-\frac{r^2}{4\beta_l t}\right).$$ [4.84]

In the preceding equations, λ is a root of the equation:

$$\frac{Q}{4\pi}e^{\lambda^2} = \lambda^2\beta_s L - \frac{k_l(T_0 - T_{\rm d})}{Ei(-\lambda^2\eta)},$$ [4.85]

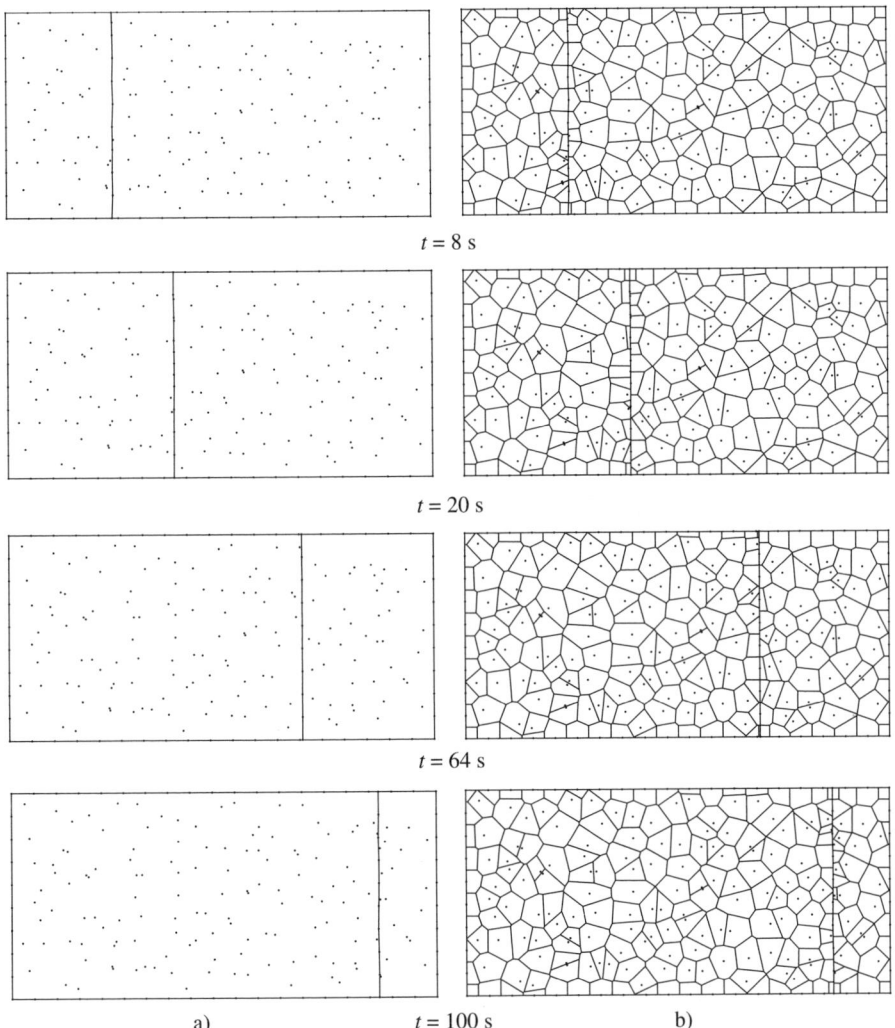

$t = 8$ s

$t = 20$ s

$t = 64$ s

a) $t = 100$ s   b)

Figure 4.35. *Position of the interface calculated by using an irregular node distribution nodes: (a) node cloud and interface position; (b) constrained Voronoi diagram*

with Q as the value of the source of heat, Ei is the exponential integral, and η the report/ratio of thermal diffusivities:

$$\eta = \frac{\beta_s}{\beta_l}. \qquad [4.86]$$

In our problem, only a quarter $\Omega = [0, 1] \times [0, 1]$ of the domain is modeled, thanks to the symmetry of the aforementioned. By imposing the exact solution on the external

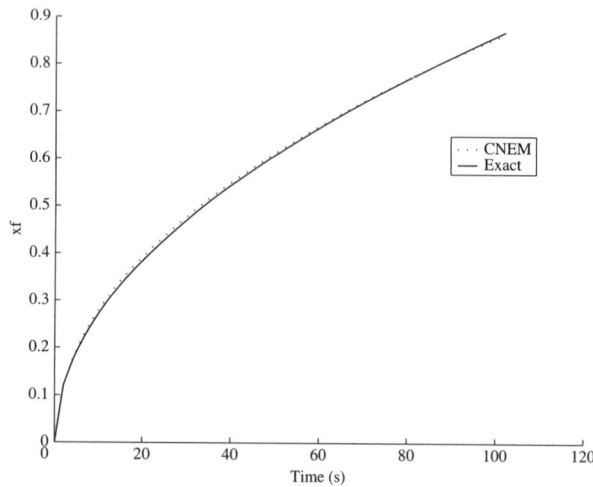

Figure 4.36. *Comparison of CNEM solution with the theoretical solution using a random grid*

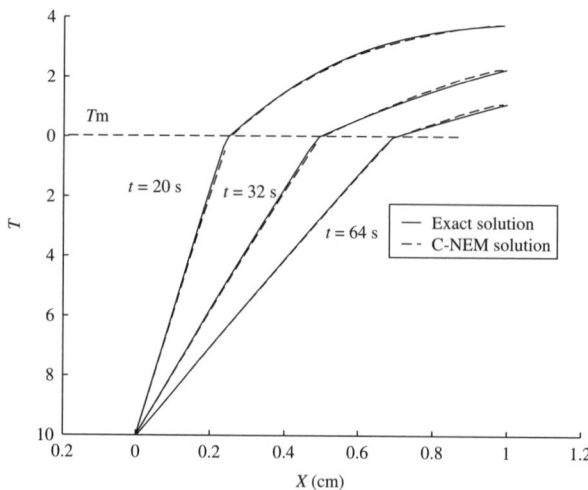

Figure 4.37. *Temperature profiles along the line $y = 0.25$ using a random grid*

boundaries, we model an axisymmetric problem in a square domain. We note that the exact solution is singular for $r = 0$. To circumvent this difficulty, we move the point located at the origin at the point $(+h/2, +h/2)$, h being the nodal distance between two nodes on the boundary. We use in the numerical applications $Q = 10$, $k_s = 9.6.10^{-2}$cal cm^{-1} s^{-1} $^0C^{-1}$, $k_l = 6.9.10^{-2}$cal cm^{-1} s^{-1} $^0C^{-1}$, and $\lambda = 0.3513$.

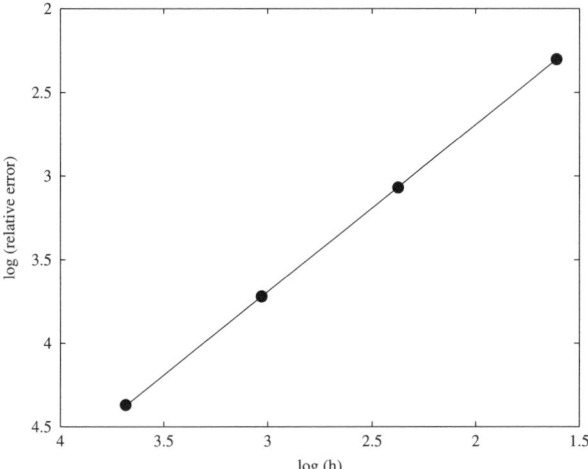

Figure 4.38. *Relative error on the average position of the front for t =50 s*

Simulations with regular and irregular grids were carried out. In the first example, the domain contains 144 nodes in a regular grid, with 8 additional nodes on the interface, initialized with its exact position for $t_0 = 1s$. Figure 4.39 shows the evolution of the constrained Voronoi cells during simulation. The Voronoi diagram as well as the CNEM shape functions are only updated locally around the interface, which reduces the calculation times compared to a total update at each time step.

In Figure 4.40, the calculated solution of the position of the front is compared with the theoretical solution. In spite of a node cloud being fixed completely at random, the form of the front remains circular during simulation. A good agreement with the analytical solution is observed, as shown in Figures 4.41 and 4.42.

Example: solidification of a casting.

In this example, the problem of solidification of a casting is studied, in which an aluminum part, initially at the temperature $T_1 = 1,000$ K, higher than the melting point $T_d = 933$ K, is cooled by suddenly subjecting it along its external boundary to a temperature of 300 K, starting the formation and the propagation of an interface between the liquid and solid phase. The geometry of the problem is represented in Figure 4.43. The problem is idealized by neglecting the influence of the formation of microstructures and by considering constant capacities and thermal conductivities, in spite of their light dependence at the temperature in the real case. The interest of this test is to study the potentialities of our approach rather than to undertake the realistic simulation of the method. The real properties of material can be found in [BRA 83]. The goal of this example is to show the capacity of the method in the propagation of the curved interfaces.

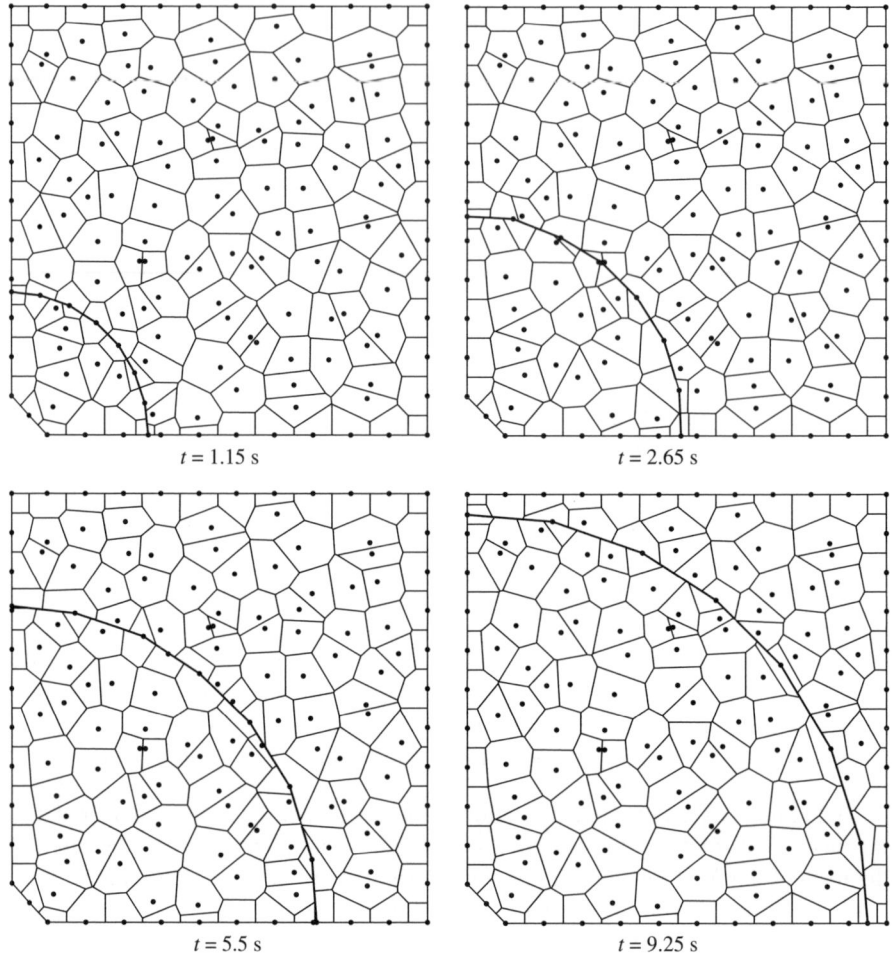

$t = 1.15$ s

$t = 2.65$ s

$t = 5.5$ s

$t = 9.25$ s

Figure 4.39. *Evolution of the constrained Voronoi cells with an irregular distribution of nodes*

In Figure 4.44, the evolution of the constrained Voronoi cells is represented for various positions of the interface. In Figure 4.45, the positions of the interface are represented at various moments.

4.6. Adiabatic shearing, cutting, and high speed blanking

In this last part, we approach the use of the CNEM in the context of large transformations (large displacements and large strain). As we see, even if a large number of functionalities are operational, certain developments are still to date under development or to come.

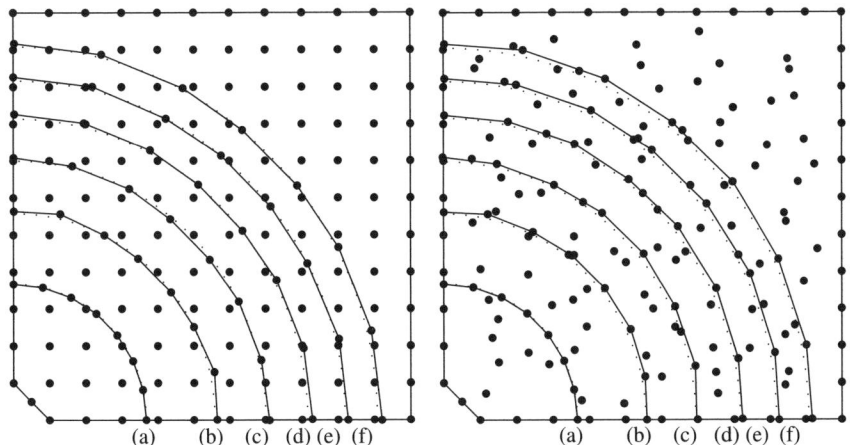

Figure 4.40. *Comparison between the calculated solution (line in continuous feature) and the analytical solution (dotted line): (a) t = 1.15 s, (b) t = 2.8 s, (c) t = 4.3 s, (d) t = 5.65 s, (e) t = 7.15 s, (f) t = 8.65 s*

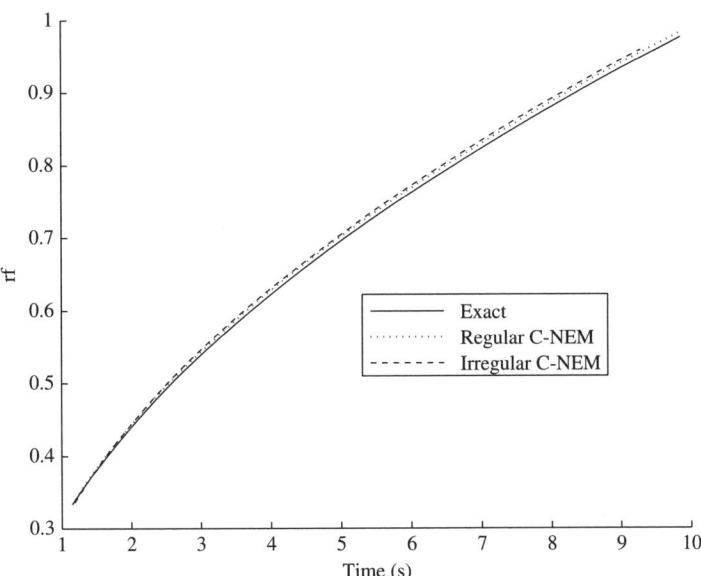

Figure 4.41. *Position of the radius of the interface*

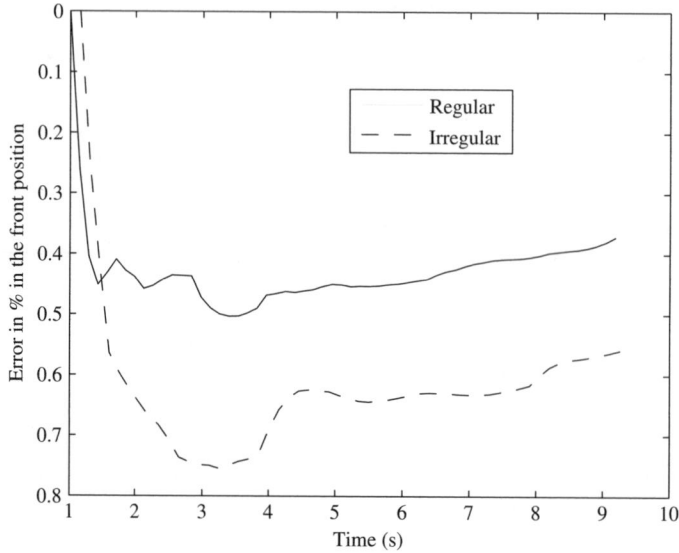

Figure 4.42. *Error in the prediction of the position of the circular front*

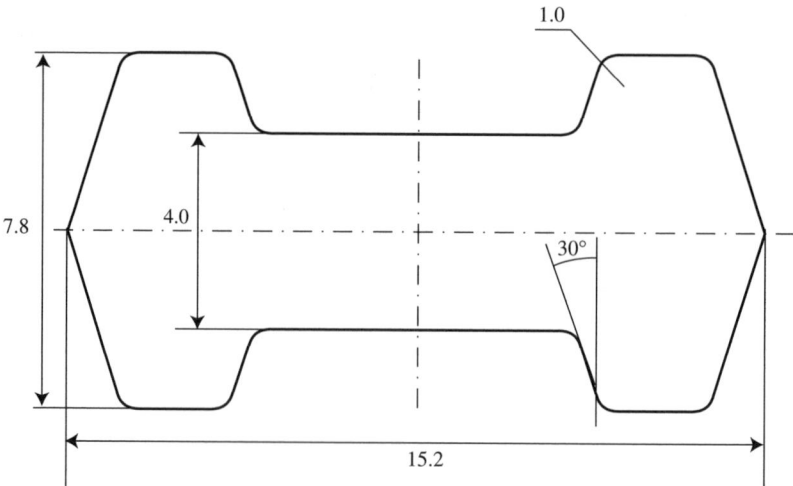

Figure 4.43. *Domain geometry*

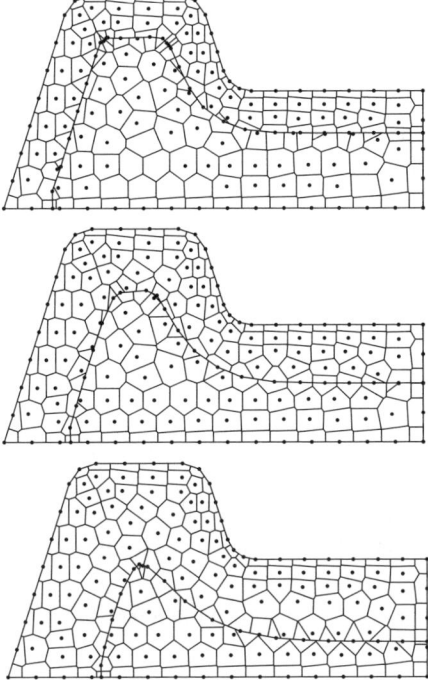

Figure 4.44. *Constrained Voronoi cells for various positions of the interface*

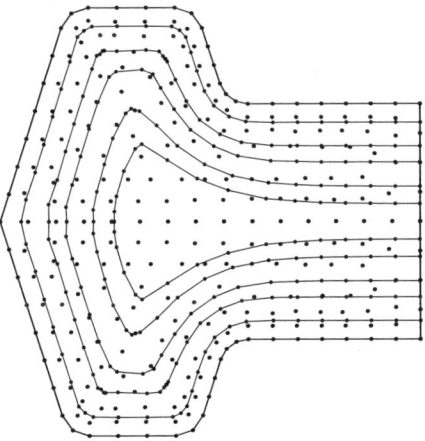

Figure 4.45. *Position of the interface at various moments*

We start by presenting the step implemented for the finite transformations: time integration scheme, characteristics related to the updating of the configuration, models behavior (model of the Johnson-Cook type), and contact. We are then interested in two particular examples. The first example is based on, the test of the bar of Taylor, which is intended to position the results obtained with the CNEM with respect to those obtained with other approaches. We also propose, in this example, a study on the influence of the thermal conductivity and of thermal softening (term of softening in the Johnson–Cook yield stress) on the results of simulation. The second example is that of high speed blanking. Two aspects are approached: the study of the first moments of shearing and the characteristic durations phenomena are encountered, then the study of the dependence of the localization of the plastic strain according to the behavior of the material (two different materials are compared).

4.6.1. *General context of the implementation of the large transformations*

The objective of this section is to specify the process, which we adopted to use the CNEM in the context of large transformations. This process contains:

– kinematic aspects:
- use of an updated Lagrangian formulation (ULF), and
- use of an explicit time integration scheme,
– contact handling: contact with friction on rigid surfaces only;
– constitutive equation integration scheme which is based on a multiplicative decomposition of the transformation gradient tensor in elastic part and plastic part.

These various points are approached one by one in the following sections. The majority of the made choices are identical to those presented in the thesis of Yvonnet [YVO 04a].

The model of Johnson–Cook is taken for the constitutive equation (equation [4.87]). This model is used extensively for the phenomena with a high strain rate because it takes into account in a simple way the main physical phenomena encountered:

– work hardening,
– dynamic hardening, and
– thermal softening.

$$\sigma\left(\varepsilon_{eq}, \dot{\varepsilon}_{eqeq}, T\right) = \underbrace{\left(A + B \cdot \left(\varepsilon_{eq}\right)^{n}\right)}_{\text{Work hardening}} \cdot \underbrace{\left(1 + C \cdot \ln\left(\frac{\dot{\varepsilon}_{eq}}{\dot{\varepsilon}_0}\right)\right)}_{\text{Dynamic hardening}} \cdot \underbrace{\left(1 - \left(\frac{T - T_t}{T_f - T_t}\right)^{M}\right)}_{\text{Thermal softening}},$$

$$[4.87]$$

where A is the stress yield to null plastic strain, B the parameter of work hardening, n the nonlinear parameter of work hardening, C the sensitivity coefficient of the strain rate, $\dot{\varepsilon}_0$ the strain rate reference, T_t the temperature of transition, T_f the melting point, and m the exponent of thermal softening.

4.6.1.1. *Updated Lagrangian formulation*

One of the main costs of the CNEM, compared to the approaches of the finite elements type relates to the construction of the interpolation and the calculation of the stabilized gradients. Because of large strains, there can be locally large displacements between neighbor nodes (in particular in the presence of shearing). Thus, the initial natural neighbors of a node may not remain, hence there is a need for reactualizing the interpolation. The displacement of a particle m at the moment t, noted as $\widetilde{\boldsymbol{u}}(m)_{0\to t}$, is then defined by

$$\widetilde{\boldsymbol{u}}(m)_{0\to t} = \widetilde{\boldsymbol{u}}(m)_{0\to t_{\text{act}}} + \widetilde{\boldsymbol{u}}(m)_{t_{\text{act}}\to t}, \qquad [4.88]$$

$$\widetilde{\boldsymbol{u}}(m)_{t_{act}\to t} = \sum_{i=1}^{\text{neighbor nb}} \phi_{i_{t_{\text{act}}}}\big(x(m)_{t_{\text{act}}}\big) \cdot \boldsymbol{u}\big(m_i\big)_{t_{\text{act}}\to t}, \qquad [4.89]$$

where t_{act} is the moment corresponding to the last updating, $x(m)_{t_{\text{act}}}$ the coordinates of the particle m at the moment t_{act}, $\phi_{i_{t_{\text{act}}}}(x)$ the shape functions built on the configuration at the moment t_{act} at the point of coordinates x, associated to neighbor i, and $\boldsymbol{u}(m_i)_{t_{\text{act}}\to t}$ the displacement associated with the ith particle close to the moment t_{act} at the moment t.

From this, the total gradient of the transformation results in a point m as defined by

$$\boldsymbol{F}_{0\to t} = \boldsymbol{F}_{t_{\text{act}}\to t} \cdot \boldsymbol{F}_{0\to t_{\text{act}}} \qquad [4.90]$$

4.6.1.2. *Processing the additional points*

As specified in section 2.3.2, the construction (or rebuilding in ULF) of the constrained Voronoi diagram requires the addition of new nodes. These new nodes are always added on the surface of the domain, and thus belong to one of the triangles describing its boundary.

These new nodes are treated as nodes "slaves" of the three nodes of the triangle on which they are located. Their displacement is defined like a linear combination of the "main" nodes.

$$\boldsymbol{u}_{\text{slave}} = \sum_{i=1}^{\text{main nb}} \lambda_i \cdot \boldsymbol{u}_i, \qquad [4.91]$$

where u_i is the displacement of each main node, and λ_p the weighting coefficient associated ($\sum_{i=1}^{\text{main nb}} \lambda_i = 1$).

This step has several advantages. It makes it possible not to add additional unknown variables to the problem to be treated and especially avoids decreasing the time step adopted in the explicit time integration scheme (*conditionally stable scheme*).

The volume of each slave cell is distributed, in proportion to λ_i, on its main nodes.

4.6.1.3. *Time integration scheme*

The time integration scheme selected is that of the central-difference [GÉR 96]. This explicit integration scheme is conditionally stable. It is classically presented in the form:

$$\ddot{u}_n = M^{-1}\big(F_n^{\text{ext}} - F_n^{\text{int}}\big), \qquad [4.92]$$

$$\dot{u}_{n+\frac{1}{2}} = \dot{u}_{n-\frac{1}{2}} + \frac{\Delta t_n + \Delta t_{n-1}}{2}\,\ddot{u}_n, \qquad [4.93]$$

$$u_{n+1} = u_n + \Delta t_n\,\dot{u}_{n+\frac{1}{2}}, \qquad [4.94]$$

where M is the mass matrix, the latter is taken diagonal (for each node, it is about the mass of the associated cell), F_n^{ext} the column of the generalized external forces, and F_n^{int} the column of the generalized internal forces.

4.6.1.4. *Processing of the contact*

Currently only the contact with friction (*friction of Coulomb*) on mobile rigid surfaces is implemented. This processing is based on a prediction/correction approach where components of F_n^{ext} is calculated so as to observe the conditions of contact at t_{n+1}.

4.6.1.5. *Integration of the constitutive equation*

For the integration of the constitutive equation, we followed the step suggested by [DE 01]. This step is based on a description of the plastic state of the matter starting from a multiplicative decomposition of the gradient of the total transformation F. This tensor is broken up into an elastic part and a plastic part:

$$F = F^e \cdot F^p. \qquad [4.95]$$

Let S_{t_n} be the configuration at the moment t_n, for which one knows σ_n the Cauchy stress tensor, F_n the gradient of the total transformation, F_n^P the plastic part of the gradient of the total transformation, and p_n the accumulated plastic strain.

And let $S_{t_{n+1}}$, be the configuration at the moment $t_{n+1} = t_n + \Delta t$.

The integration of the constitutive equation consists of finding σ_{n+1}, F^P_{n+1}, and p_{n+1} giving only the gradient of the total transformation F_{n+1} to the moment t_{n+1}.

The time integration scheme makes it possible to calculate F_{n+1} on each node.

$$F_{n+1} = I + \nabla_{S_0}(u_{n+1}). \qquad [4.96]$$

In the context of ULF $F_{n+1} = (I + \nabla_{S_{t_{act}}}(u_{t_{act} \to t_{n+1}})) \cdot F_{0 \to t_{act}}$.

The integration of the constitutive equation begin with an elastic prediction which consists in making the assumption that $F^p_{n+1} = F^p_n$.

Elastic prediction

The predictive elastic gradient $^*F^e_{n+1}$ is defined by

$$^*F^e_{n+1} = F_{n+1} \cdot (F^p_n)^{-1} = {}^*R^e \cdot {}^*U^e, \qquad [4.97]$$

where $^*R^e$ and $^*U^e$ are, respectively, the elastic right stretch tensor and the elastic rotation.

The predictive turned Cauchy stress tensor $^*\hat{\sigma}$ is defined by the elastic constitutive equation written in equations [4.98] and [4.99].

$$\mathrm{Tr}(^*\hat{\sigma}) = 3\,\kappa \mathrm{Tr}(^*E^e), \mathrm{dev}(^*\hat{\sigma}) = 2\,\mu \mathrm{dev}(^*E^e)\, and \qquad [4.98]$$

$$^*\hat{\sigma} = J \cdot (^*R^e)^T \cdot {}^*\sigma \cdot {}^*R^e, \text{ with } J = \det(^*F^e_{n+1}) = \det(F_{n+1}), \qquad [4.99]$$

where $^*E^e = ln(^*U^e)$ is the logarithmic strain tensor of Henky, $3\,\kappa = \frac{E}{1-2\nu}$ the hydrostatic compression modulus (E Young modulus, ν Poisson's ratio), and $2\,\mu = \frac{E}{1+\nu}$ the transverse stiffness modulus.

The equivalent predictive stress $^*\sigma^{eq}$ (*Von Mises equivalent stress*) is worth as follows:

$$^*\sigma^{eq} = \frac{1}{J} {}^*\hat{\sigma}^{eq} = \frac{1}{J}\sqrt{\frac{3}{2}\mathrm{dev}(^*\hat{\sigma}):\mathrm{dev}(^*\hat{\sigma})}. \qquad [4.100]$$

Thereby noting σ^y_n as the yield stress at the beginning of increment, two cases are to be considered:

– $^*\sigma^{eq} \leq \sigma^y_n$ → one stays in elastic domain, the prediction is the solution.

– $^*\sigma^{eq} \leq \sigma^y_n$ → one is in the plastic domain, it is thus necessary to apply the following plastic correction.

Plastic correction

It is supposed that the plastic flow is done in the direction of the stress deviator tensor (*normality law*):

$$\hat{\boldsymbol{D}}^p = \dot{p}\,\frac{3}{2}\,\frac{\mathrm{dev}(\hat{\boldsymbol{\sigma}})}{\hat{\sigma}^{eq}}, \qquad [4.101]$$

and thus, by rear Euler integration (*implicit*), that:

$$\Delta\boldsymbol{E}^p = \Delta p\,\frac{3}{2}\,\frac{\mathrm{dev}(\hat{\boldsymbol{\sigma}})}{\hat{\sigma}^{eq}}, \qquad [4.102]$$

from which the elastic strain is deduced $\boldsymbol{E}^e$:

$$\boldsymbol{E}^e = {}^*\boldsymbol{E}^e - \Delta\boldsymbol{E}^p. \qquad [4.103]$$

From the elastic strain, it is possible to calculate the turned stress deviator tensor $\mathrm{dev}(\hat{\sigma})$:

$$\mathrm{dev}(\hat{\boldsymbol{\sigma}}) = 2\,\mu\mathrm{dev}(\boldsymbol{E}^e) = 2\,\mu\mathrm{dev}({}^*\boldsymbol{E}^e) - 2\,\mu\,\Delta\mathrm{dev}(\boldsymbol{E}^p), \qquad [4.104]$$

or:

$$\mathrm{dev}(\hat{\boldsymbol{\sigma}}) = \mathrm{dev}({}^*\hat{\boldsymbol{\sigma}}) - 3\,\mu\,\Delta p\,\frac{\mathrm{dev}(\hat{\boldsymbol{\sigma}})}{\hat{\sigma}^{eq}}. \qquad [4.105]$$

That is to say, finally

$$\hat{\sigma}^{eq} = {}^*\hat{\sigma}^{eq} - 3\,\mu\,\Delta p. \qquad [4.106]$$

Thus, the plastic flow taking place is given in equation [4.107].

$$\hat{\sigma}^{eq} = \sigma^y(p_n, \Delta p, \dots). \qquad [4.107]$$

By replacing the expression $\hat{\sigma}^{eq}$ given by equation [4.107] in equation [4.106], we obtain a nonlinear equation from which the solution makes it possible to determine the increment of accumulated plastic stress Δp. Once Δp is known, equation [4.105] gives us $\mathrm{dev}(\hat{\boldsymbol{\sigma}})$.

Considering the plastic flow is incompressible ($\det \boldsymbol{F}^p = 1$), $\mathrm{Tr}(\boldsymbol{E}^e) = \mathrm{Tr}({}^*\boldsymbol{E}^e)$ and thus

$$\mathrm{Tr}(\hat{\sigma}) = 3\,\kappa\mathrm{Tr}(\boldsymbol{E}^e) = 3\,\kappa\mathrm{Tr}({}^*\boldsymbol{E}^e), \qquad [4.108]$$

hence

$$\hat{\boldsymbol{\sigma}}_{n+1} = \mathrm{dev}(\hat{\sigma}) + \frac{1}{3}\mathrm{Tr}(\hat{\sigma}). \qquad [4.109]$$

To go back to $\boldsymbol{F}^p_{n+1}$ it is enough to follow the procedure:

$$\boldsymbol{U}^e = \exp(\boldsymbol{E}^e) \longrightarrow \boldsymbol{F}^e_{n+1} = {}^*\boldsymbol{R}^e \cdot \boldsymbol{U}^e \longrightarrow \boldsymbol{F}^p_{n+1} = \left(\boldsymbol{F}^e_{n+1}\right)^{-1} \cdot \boldsymbol{F}_{n+1}. \quad [4.110]$$

4.6.2. *Applications*

4.6.2.1. *Taylor's bar*

The Taylor's bar test consists of impacting a (cylindrical) metal bar with an uniform initial velocity v_c on a rigid plane surface (Figure 4.46). For our simulations, the axis of the cylinder as well as the initial velocity of the cylinder, is taken parallel with the normal of the impacted plane surface.

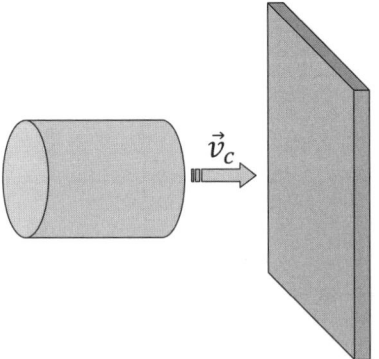

Figure 4.46. *Diagram of the test of Taylor's bar*

Data, experimental results, as well as the results of finite elements simulations (LS-Dyna, with thermomechanical coupling) of this test were taken again from [BAN 05].

The data used are the following:

– Material: Cu-ETP (*Cu-a1*) whose behavior is modeled via a model of Johnson–Cook type (parameters are given in Table 4.4).

– Dimensions:
  - initial length: 30 mm,
  - initial diameter: 6 mm.

– Impact velocity: 188 m/s.

– Initial temperature: 727 °K.

– Contact made without friction and impacted plane surface supposed infinitely rigid.

A (MPa)	B (MPa)	C	n	m	$\dot{\varepsilon}_{p0}$ (s^{-1})	T_t (°K)	T_f (°K)
90	292	0.025	0.31	1.09	1	294	1,356

Table 4.4. *Data of the material: Cu-ETP (Cu-a1)*

In Figure 4.47, the profiles of the cylinder after impact are represented for: the experimental result, a finite element simulation carried out with LS-Dyna, and finally, the simulation carried out with the CNEM. For the two simulations, thermal conductivity was taken into account.

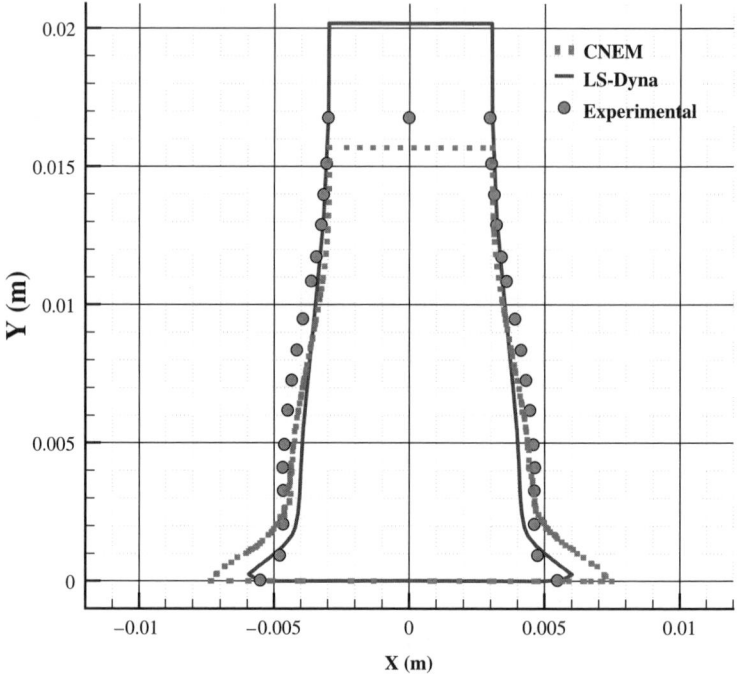

Figure 4.47. *Comparison between the profiles obtained: FEM/CNEM experiment*

The final lengths and diameters are listed in Table 4.5.

	Experimental	LS-Dyna	CNEM
Final length (mm)	16.8	20.2	15.7
$(l - l_{exp})/l_{exp}$	—	20.5%	−6.3%
Final diameter (mm)	11.0	12.0	14.8
$(d - d_{exp})/d_{exp}$	—	8.8%	34.7%

Table 4.5. *Dimensions of Taylor's bar after impact*

The various stages of the crushing of a cylinder are shown in Figures 4.49 – 4.54. Two great phases are identified: the first phase during which only the end widens, and the second phase when a bulge in the central part appears. At the end of the simulation ($t = 1, 18.10^{-4}\,s$), we observe a relaxation of the stresses: only the residual stresses remain to which elastic waves are added.

The CNEM and finite elements results seem to circle the experimental result (the CNEM gives a better prediction over the length, and the finite elements a better prediction on the diameter). However, it is worth noting that these simulations do not integrate the cylindrical/front friction, which tend to reduce the diameter and increase the length of the cylinder after impact.

To study the influence of the thermal phenomena on the result obtained, the same simulation was carried out:

– adiabatically and

– with a constant temperature (no thermal softening).

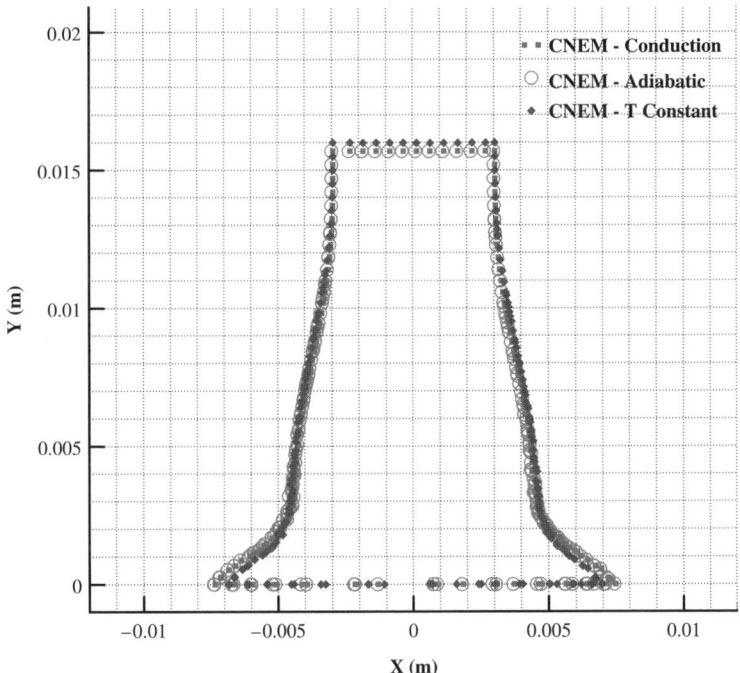

Figure 4.48. *Comparison between profiles obtained: conduction/adiabatic/ T constant*

The profiles of the cylinder obtained are presented in Figure 4.48. The lengths as well as the final diameters are given in Table 4.6.

	Conduction	Adiabatic	T constant
Final length (mm)	15.7	15.7	16.0
Final diameter (mm)	14.8	14.8	13.8
Max T(°K)	960	980	718
Max p	3.41	3.41	2.93

Table 4.6. *Taylor's bar: results according to the design criteria*

Regarding the final profile of the cylinder, few differences are observed for three simulations. Only the suppression of thermal softening has a very light influence on the result (less crushing); this difference is, however, more significant for the temperatures (and cumulated plastic strain): 718 °C for the case without softening against 960 or 980 °C for two other simulations.

4.6.2.2. *Adiabatic shearing*

An experimental device of high speed blanking was put into a diagram, as indicated in Figure 4.55. This device comprises a punch which, launched with an

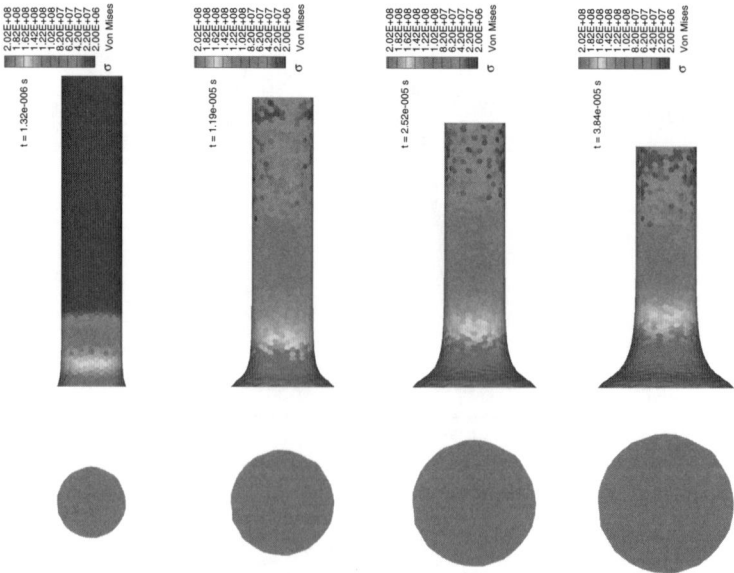

Figure 4.49. *Taylor's bar – equivalent Stress of Von Mises*
$-0s < t < 3,84.10^{-5}s$

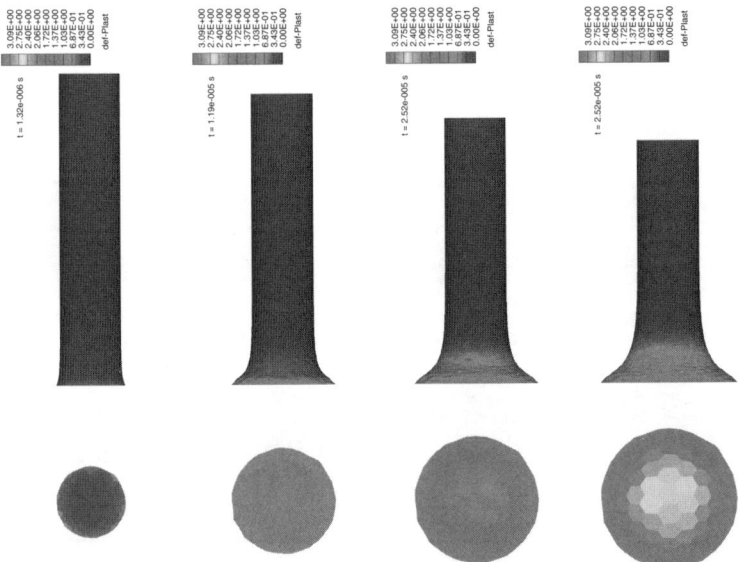

Figure 4.50. *Taylor's bar – cumulated plastic strain $-0s < t < 3,84.10^{-5}s$*

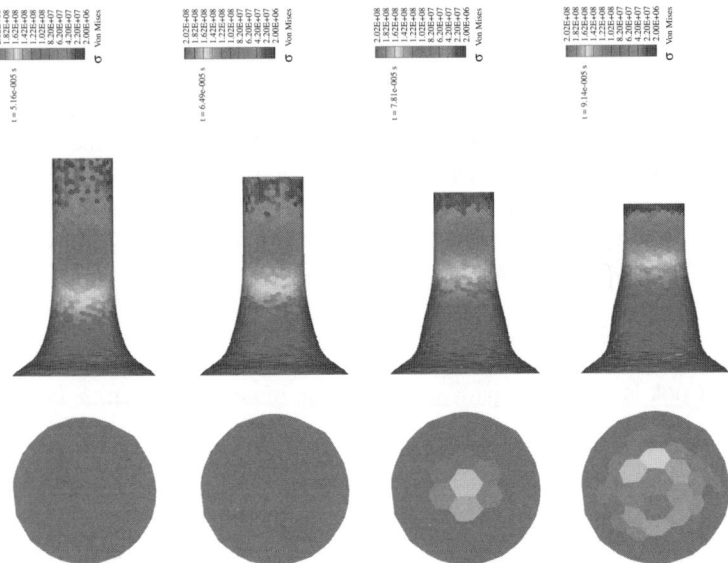

Figure 4.51. *Taylor's bar – equivalent Stress of Von Mises*
$-3.84 \times 10^{-5}s < t < 9.14 \times 10^{-5}s$

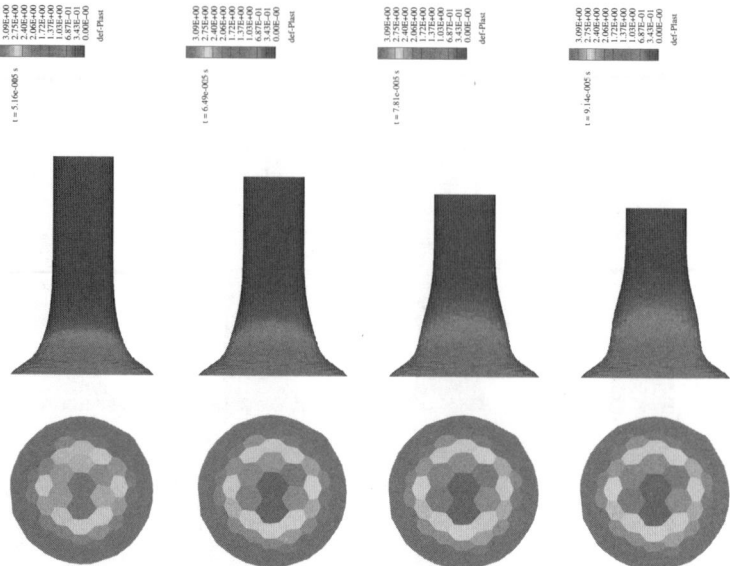

Figure 4.52. *Taylor's bar – cumulated plastic strain*
$-3.84 \times 10^{-5}s < t < 9.14 \times 10^{-5}s$

initial velocity v_p, comes to impact a parallelepipedic workpiece simply posed on a matrix, itself installed on a Hopkinson tube (tube not represented on the figure).

For simulations, the punch and the die are considered rigid, the die being fixed and the punch being supposed to operate with a constant speed. As the matrix-punch-workpiece unit has two plane symmetries, only a quarter of the device is modeled. Thermal conductivity was not taken into account. The contacts punch/workpiece and workpiece/die are presumed frictionless. The procedure for updating the configuration was not applied to this example.

In the following section, we have the results for two different materials: stainless steel, the 304 L, and an alloy of titanium, the TA6V. In both cases, the model of Johnson–Cook is used. The parameters of these models are given in Tables 4.7 and 4.8.

A (MPa)	B (MPa)	C	n	m	$\dot{\varepsilon}_{p0}(s^{-1})$	T_t (°K)	T_f (°K)
253.32	685.1	0.097	0.3128	2.044	1	296	1,698

Table 4.7. *Data of material-a: 304 L*

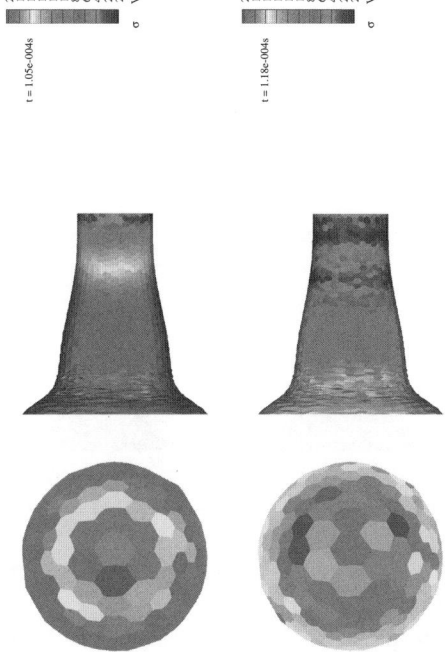

Figure 4.53. *Taylor's bar – equivalent stress of Von Mises*
$-9.14 \times 10^{-5} < t < 1,18.10^{-4}$

A (MPa)	B (MPa)	C	n	m	$\dot{\varepsilon}_{p0}\ (s^{-1})$	$T_t\ (°C)$	$T_f\ (°C)$
866	318	0.008	0.25	1.055	5.77×10^{-2}	20	1,670

Table 4.8. *Data of material-b (according to [RAN 04]): TA6V*

The presented simulations of shearing have the following data in common:

– gap between the punch and the die j: 0.3 mm,

– width of the punch l_p: 11.4 mm,

– edge radius of the punch r_p: 0.2 mm,

– edge radius of the die r_m: 0.2 mm,

– thickness of workpiece e_e: 5 mm,

– width of workpiece l_e: 8 mm,

– length of workpiece h_e: 24 mm,

– speed of the punch: 10 m/s, and

– initial temperature: 293 °K.

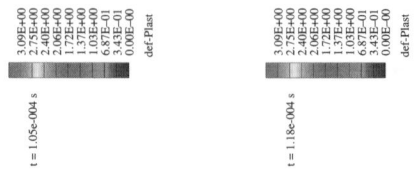

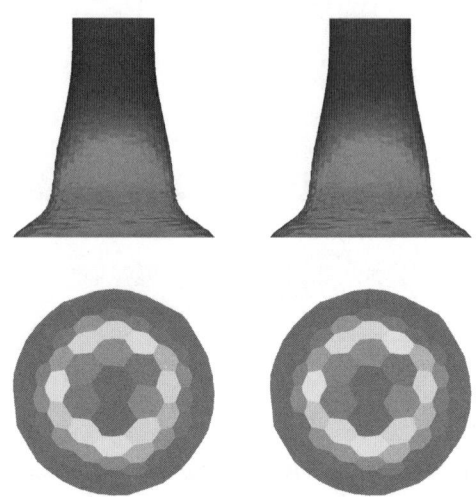

Figure 4.54. *Taylor's bar – cumulated plastic strain –*
$9.14 \times 10^{-5} < t < 1,18.10^{-4}$

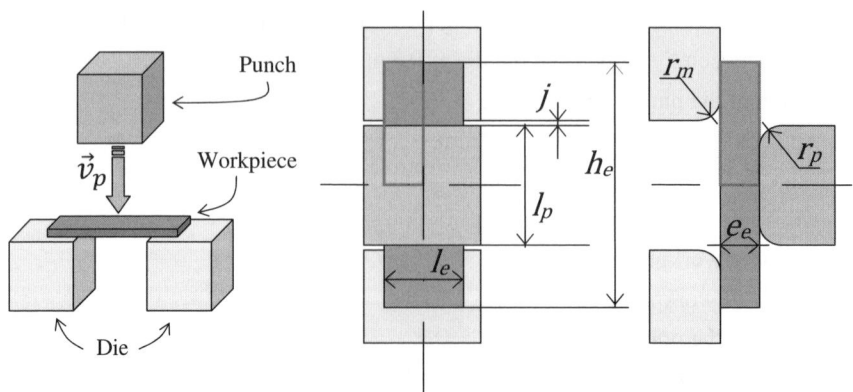

Figure 4.55. *Diagram representing the adiabatic shearing device*

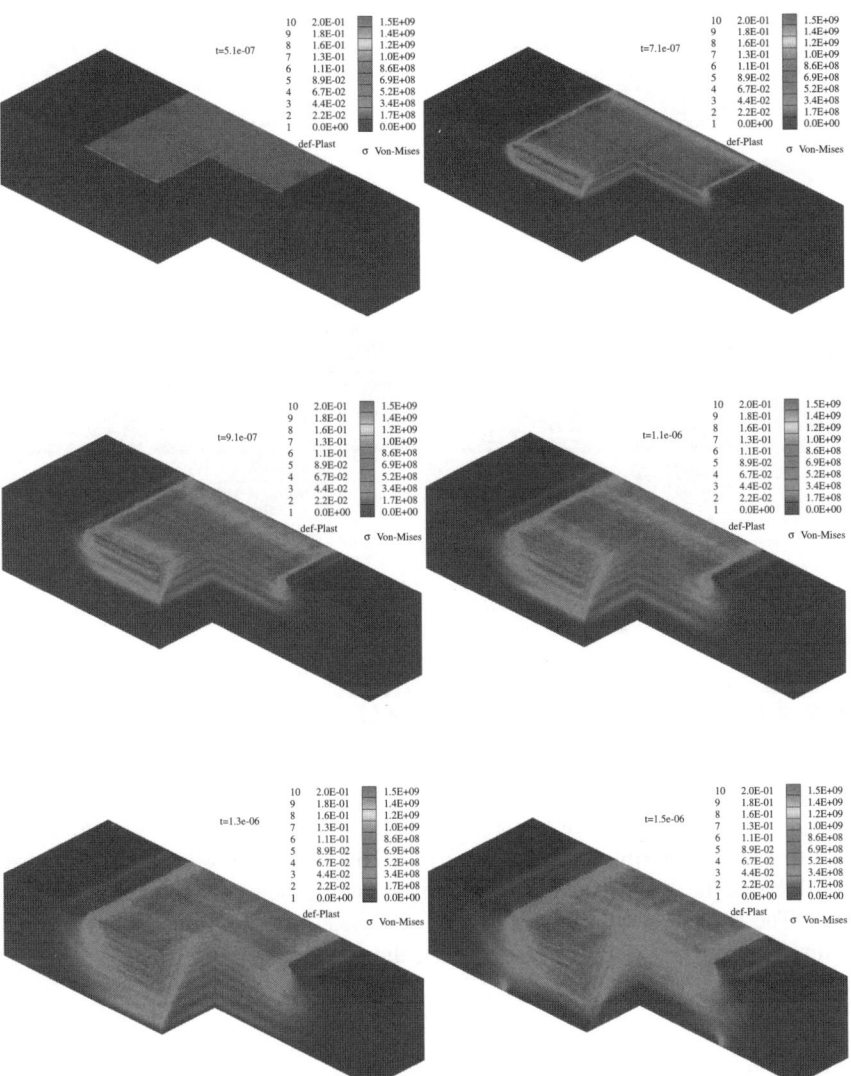

Figure 4.56. *Shearing 304 L −0s < t < 1.5 × 10⁻⁶s*

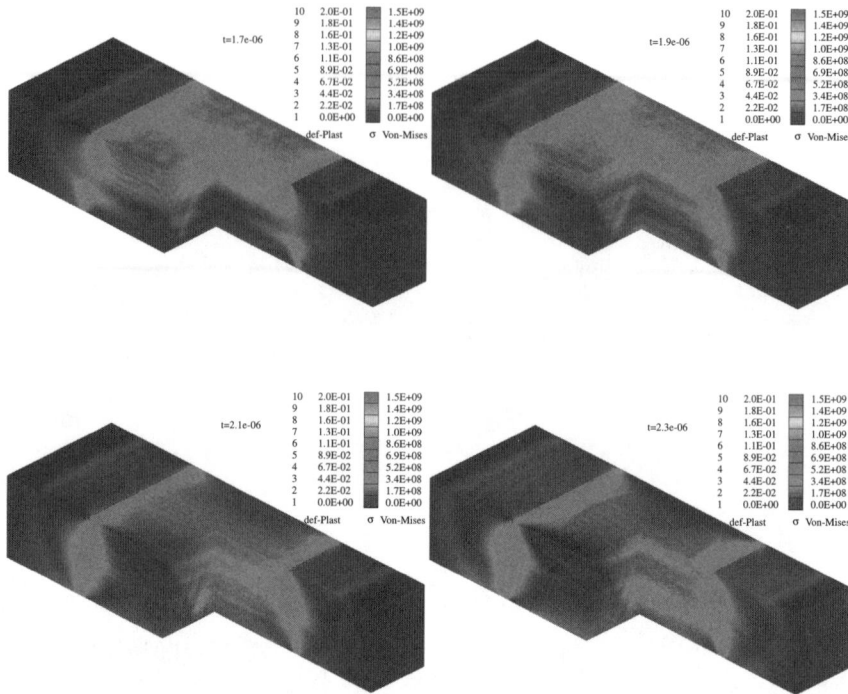

Figure 4.57. *Shearing 304 L $-1.5 \times 10^{-6}s < t < 2.3 \times 10^{-6}s$*

Study of the beginning of shearing (steel 304 L)

In Figures 4.56 and 4.57, the first moments of shearing are shown ($0 \leqslant t \leqslant 2,3.10^{-6}s$). The elastic wave propagation is clearly visible: the waves approximately cross the workpiece (thickness 5 mm) in 1×10^{-6} s. An outward return thus takes 2×10^{-6} s, at the end of which we can clearly observe the localization of the shearing areas. The level of stress reached explain that plastification in this area has already begun, showing that shearing started even though the dynamic balance is very far from being established.

It is, however, important to note that for this simulation the punch and the die are presumed to be rigid. In practice, Young's modulus and the density of the latter are close to those of a sheared material (stainless 304 L). The elastic behavior of the die and the punch cannot thus be neglected during the first moments of cutting. Considering them rigid, an overestimation of a factor of approximately two is induced on the speed of the workpiece/punch interface, and thus of the stresses present in the elastic waves.

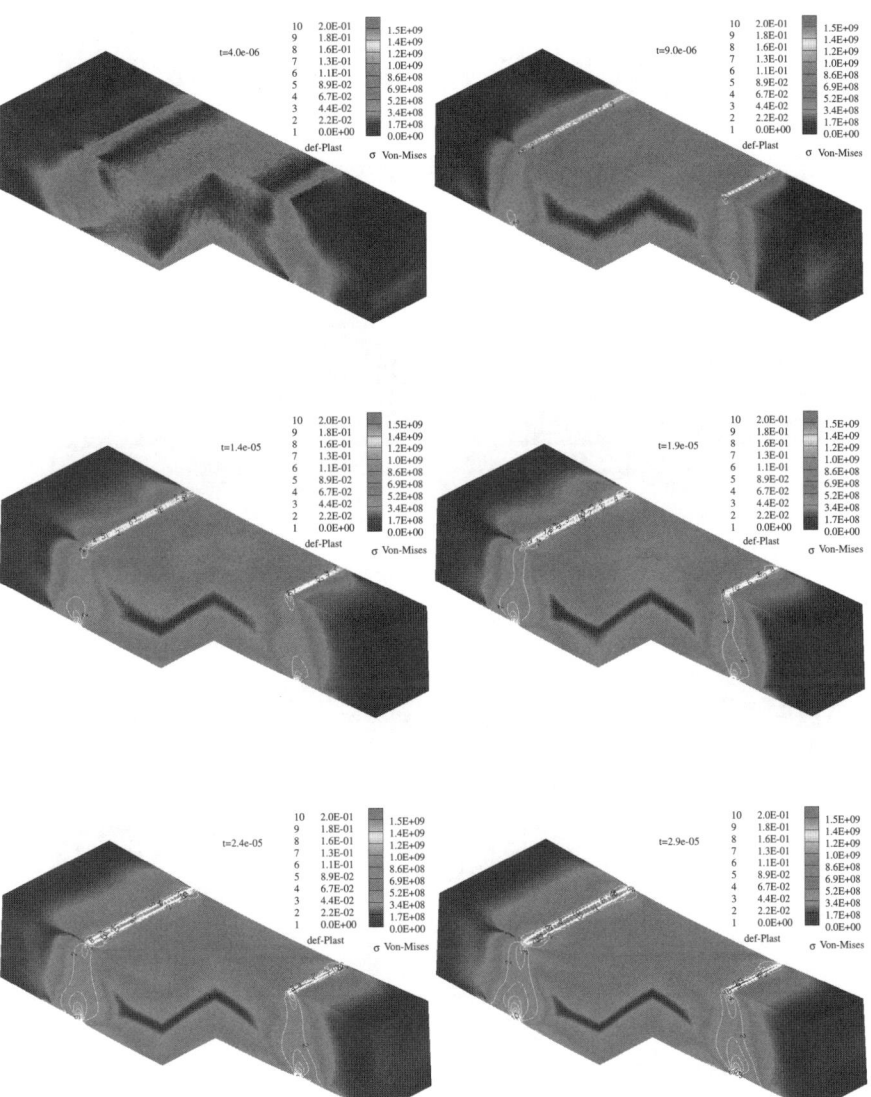

Figure 4.58. *Shearing 304 L −2.3 × 10⁻⁶ s < t < 2.9 × 10⁻⁵ s*

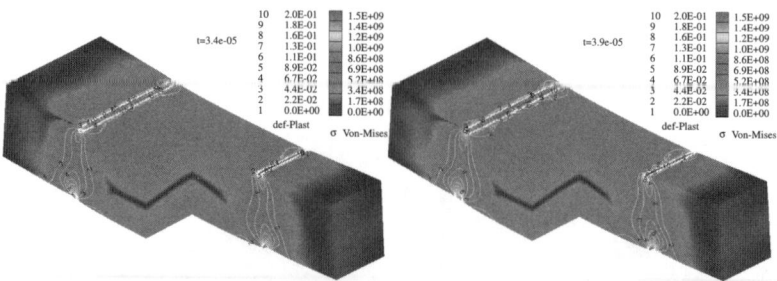

Figure 4.59. *Shearing 304 L* $-2.9 \times 10^{-5}s < t < 3.9 \times 10^{-5}s$

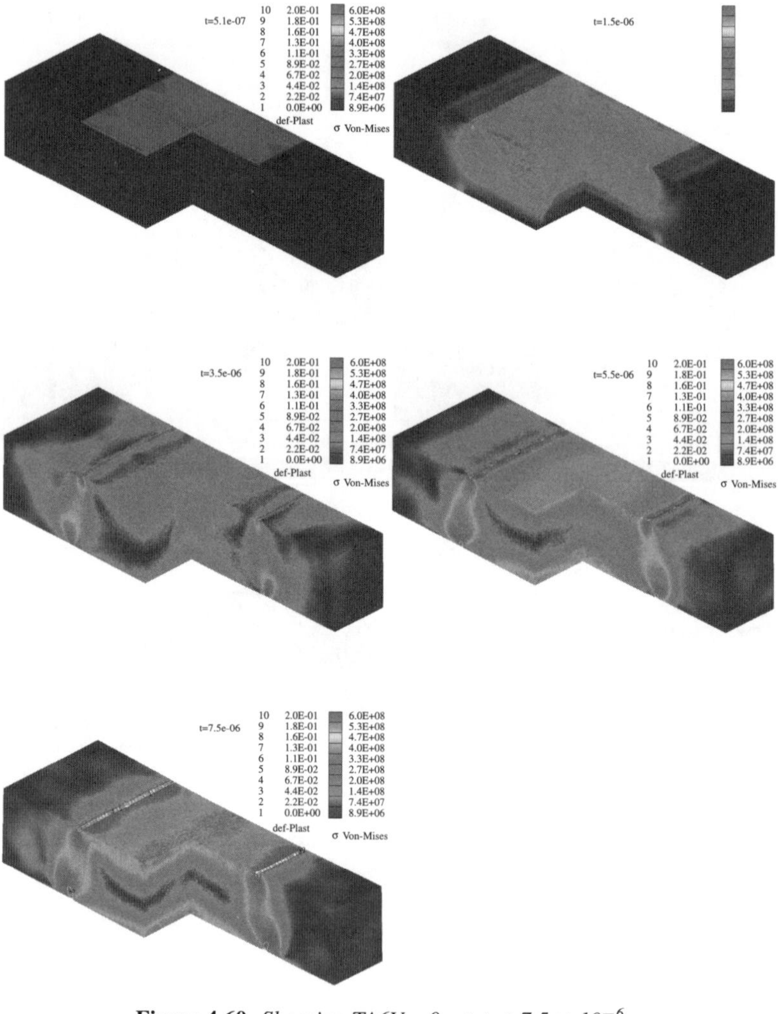

Figure 4.60. *Shearing TA6V* $-0s < t < 7.5 \times 10^{-6}s$

The continuation of simulation confirms the localization observed during the first moments (Figures 4.58 and 4.59). In its current version, the simulation, following the absence of readjustment of the domain boudnary, cannot exceed $39 \times 10^{-6}s$. This duration corresponds to approximately 20% of the duration taken for complete shearing (according to an experimental estimate).

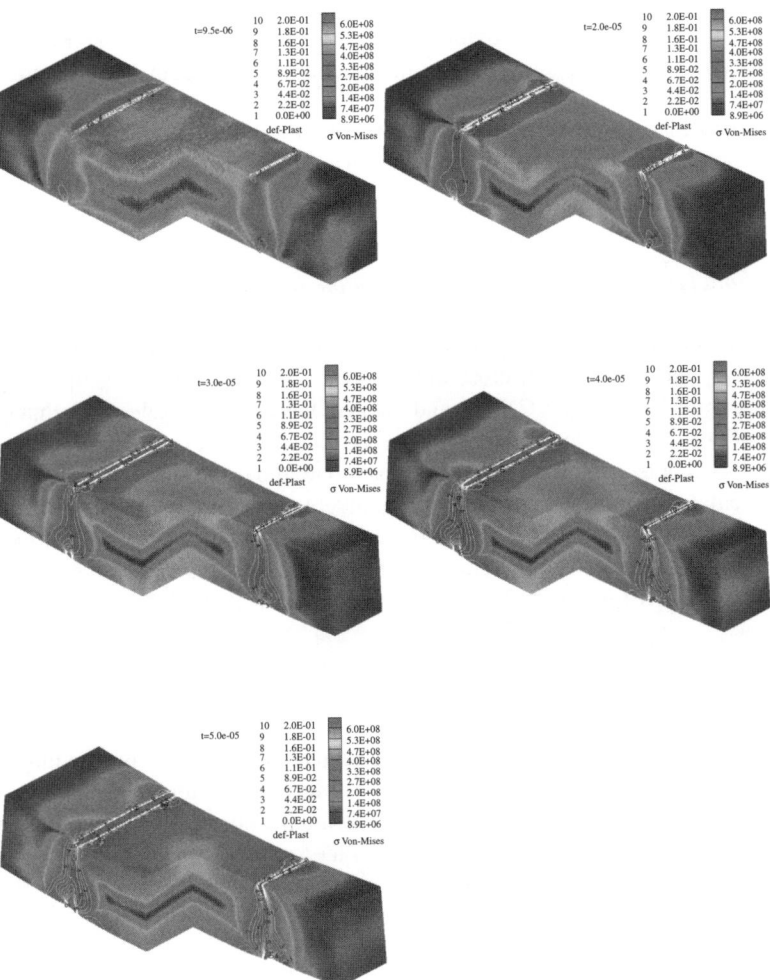

Figure 4.61. *Shearing TA6V $-7.5 \times 10^{-6}s < t < 5 \times 10^{-5}s$*

Comparative study of the localization of shearing–304 L/TA6V

We also carried out a simulation for TA6V (Figures 4.60 and 4.61). This material has a substantially different behavior from 304 L. As these figures show, simulations make it possible to find what is observed in experiments, knowing a greater localization of the shear band for TA6V. For 304L, a strong concentration of the stresses is observed in the neighborhood of the edges of the punch and the die: there is a factor of two on the stresses between these areas and the center of the shear band. For TA6V, the stress is almost homogeneous throughout the shear band.

It is necessary, however, to keep in mind that these results greatly depend on the identification of the parameters used in the constitutive equation. They simply indicate a tendency.

4.6.3. Conclusions on the numerical simulation of shearing

The simulations carried out on the Taylor's bar enabled us to compare the results obtained by the CNEM with those obtained by LS-Dyna. These results depend on the algorithms used in the CNEM and also on the model of selected behavior and the identification of its parameters. These various aspects explain, without which it is possible to determine the influence of each one, the differences observed between experimental and numerical results. The main conclusion is that the CNEM, and the implementation that we made of it, make it possible to obtain results, for this example, of quality comparable with the concurrent approaches. Finally, always for Taylor's bar, and a velocity impact of 188 m/s, simulations show that it is valid to be placed in adiabatic.

The simulations carried out in the example of blanking are currently limited to the first moments of shearing. In spite of the localization of the strain as already observed clearly, two functionalities are still lacking which prevent us in proceeding further with simulation: the possibility of adding nodes during the simulation, and separating the matter. The first feature is in the phase of finalization. The second feature may be based on a kill-cell technique which is similar to the kill-element technique. With this type of strategy being relatively simple to integrate, we hope to complete shearing simulations very soon.

Chapter 5

A Mixed Approach to the Natural Elements

5.1. Introduction

In Chapter 2, the natural elements method was developed based on the virtual work principle.

As a conclusion, for a linear two-dimensional elastic solid occupying a domain A, we obtain the following stages.

Virtual work principle:

$$\int_A \sigma_{ij} \delta\varepsilon_{ij} \, \mathrm{d}A - \int_A F_i \delta u_i \, \mathrm{d}A - \int_{S_t} T_i \delta u_i \, \mathrm{d}S = 0,$$

where $\delta\varepsilon_{ij}$ are the virtual strains associated with the virtual displacements δu_i:

$$\delta\varepsilon_{ij} = \frac{1}{2}\left(\frac{\partial \delta u_i}{\partial X_j} + \frac{\partial \delta u_j}{\partial X_i}\right);$$

where σ_{ij} are the Cauchy stresses deduced from the real displacements u_i by Hooke's law:

$$\sigma_{ij} = C_{ijkl}\varepsilon_{kl},$$

with

$$\varepsilon_{kl} = \frac{1}{2}\left(\frac{\partial u_k}{\partial X_l} + \frac{\partial u_l}{\partial X_k}\right).$$

T_i are the surface tractions and F_i the body forces applied to the solid. S_t is the part of the boundary where the surface tractions are applied. $\tilde{u}_i$ displacements can be imposed on the S_u part of the boundary of the solid.

Interpolation of displacements fields:

$$u_i = \sum_{J=1}^{N} \Phi_J u_i^J,$$

where N is the number of nodes distributed in the domain and on its boundary, $\Phi_J(X_1, X_2)$ are the Laplace or Sibson interpolation functions, and u_i^J are the nodal displacements.

Discretized virtual work principle

$$\sum_{J=1}^{N} \delta u_i^J \int_A \sigma_{ij} B_j^J \, \mathrm{d}A - \sum_{J=1}^{N} \delta u_i^J P_i^J = 0.$$

Successively, we can deduce that

$$\int_A \sigma_{ij} B_j^J \, \mathrm{d}A = P_i^J, \qquad \int_A C_{ijkl} \varepsilon_{kl} B_j^J \, \mathrm{d}A = P_i^J,$$

$$\sum_{I=1}^{N} \left[\int_A C_{ijkl} B_l^I B_j^J \, \mathrm{d}A \right] u_k^I = P_i^J, \qquad \sum_{I=1}^{N} K_{ik}^{IJ} u_k^I = P_i^J,$$

where K_{ik}^{IJ} is the stiffness matrix of the discretized solid,

$$K_{ik}^{IJ} = \int_A C_{ijkl} B_l^I B_j^J \, \mathrm{d}A.$$

and P_i^J are the nodal forces that are energetically equivalent to the surface tractions and body forces:

$$P_i^J = \int_A F_i \Phi_J \, \mathrm{d}A + \int_{S_t} T_i \Phi_J \, \mathrm{d}S.$$

and B_j^J is the matrix connecting the virtual strain to the virtual displacements:

$$\delta \varepsilon_{ij} = \sum_{J=1}^{N} B_j^J \delta u_i^J.$$

This matrix contains the derivatives of the interpolation function.

Using the Laplace or Sibson interpolation functions it is possible to impose boundary conditions of the type $u_i = \tilde{u}_i$ on S_u.

For the numerical integration of the stiffness matrix terms, a common practice consists of dividing the domain A into simple shape subdomains (triangles or quadrilaterals) in which we apply classical methods of numerical integration.

More advanced methods discussed in [ATL 99, ATL 00, CUE 03] consider the shape of the interpolation function support. In general, these methods do not give excellent results for the patch test [CUE 03].

However, we can use the stabilized integration [CHE 01, YOO 04] whose fundamental idea consists of evaluating average strains ε_{kl} in the vicinity of node I on the Voronoi cell A_I associated with this node, is given as follows:

$$\bar{\varepsilon}_{kl} = \frac{1}{A_I} \int_{A_I} \varepsilon_{kl} \mathrm{d}A_I = \frac{1}{2A_I} \oint_{C_I} \left(N_k u_l + N_l u_k \right) \mathrm{d}C_I \bar{\varepsilon}_{kl},$$

where C_I is the contour of the Voronoi cell A_I and $N = (N_1, N_2)$ is the unit normal exterior to C_I.

This concept was applied successfully [GON 04a, YVO 04b, YOO 04, CUE 03, YVO 04a], and excellent results for the patch test were obtained.

In particular, in [YVO 04b], the stresses in the Voronoi cells are deduced from the average strains, and the application of the divergence theory makes it possible to avoid numerical integration on the surface area of this cell by replacing it with a contour integration of this cell.

The aim of this chapter is to offer an alternative method to the natural element method on the basis of the Fraeijs de Veubeke variational principle [VEU 51], which allows more flexibility in the discretization of the various mechanical fields. The developments that follow are based on the works of [CES 06, CES 07, CES 08, LI 07, LI 08].

5.2. The Fraeijs de Veubeke variational principle for linear elastic problems

In 1954, Hu Hai Chang [HU 55] proposed the following functional for linear elastic solids:

$$\Pi = \int_A L_U \, \mathrm{d}A - \int_{S_u} N_j \Sigma_{ji} \tilde{u}_i \, \mathrm{d}S - \int_{S_t} u_i \left(N_j \Sigma_{ji} - T_i \right) \mathrm{d}S, \qquad [5.1]$$

with

$$L_U = \frac{\partial \Sigma_{ij}}{\partial X_j} u_i + F_i u_i + \varepsilon_{ij} \Sigma_{ij} - W(\varepsilon_{ij}), \qquad [5.2]$$

where A is the area of the domain, S the domain contour, N_j the unit normal exterior to this contour, S_t and S_u the parts of S on which surface tractions T_i and $\tilde{u}_i$ displacements are, respectively, imposed.

This functional has three fields:

$$\Pi = \Pi(u_i, \Sigma_{ij}, \varepsilon_{ij})$$

– the displacement field (u_i),
– the stress field (Σ_{ij}), and
– the strain field (ε_{ij}).

$W(\varepsilon_{ij})$ is the strain energy density from which we can deduce the "constitutive stresses":

$$\sigma_{ij} = \frac{\partial W}{\partial \varepsilon_{ij}}. \qquad [5.3]$$

We can easily verify that the functional present in equation [5.1] can be written in the following way:

$$\Pi = \int_A W(\varepsilon_{ij}) \, \mathrm{d}A + \int_A \Sigma_{ij} \left[\frac{1}{2} \left(\frac{\partial u_i}{\partial X_j} + \frac{\partial u_j}{\partial X_i} \right) - \varepsilon_{ij} \right] \mathrm{d}A$$
$$- \int_A F_i u_i \, \mathrm{d}A - \int_{S_t} T_i u_i \, \mathrm{d}S + \int_{S_u} N_j \Sigma_{ji} (\tilde{u}_i - u_i) \, \mathrm{d}S. \qquad [5.4]$$

This functional was proposed independently by Washizu [WAS 55] in 1955, and now, it is known as the Hu–Washizu functional. However, an article published in 2000 by Felippa [FEL 00] showed that this functional was proposed in 1951 by Fraeijs de Veubeke [VEU 51] in a more general form, which can be written as follows:

$$\Pi = \Pi(u_i, \Sigma_{ij}, \varepsilon_{ij}, r_i) = \int_A W(\varepsilon_{ij}) \mathrm{d}A + \int_A \Sigma_{ij} \left[\frac{1}{2} \left(\frac{\partial u_i}{\partial X_j} + \frac{\partial u_j}{\partial X_i} \right) - \varepsilon_{ij} \right] \mathrm{d}A$$
$$- \int_A F_i u_i \, \mathrm{d}A - \int_{S_t} T_i u_i \, \mathrm{d}S + \int_{S_u} r_i (\tilde{u}_i - u_i) \, \mathrm{d}S. \qquad [5.5]$$

It has an additional independent field r_i, which is identified with the support reactions on S_u.

If we assume that these reactions are a priori in equilibrium with the assumed stresses Σ_{ij}: $r_i = N_j \Sigma_{ji}$, we find the Hu–Washizu functional.

The variation of $\Pi = \Pi(u_i, \Sigma_{ij}, \varepsilon_{ij}, r_i)$ can be written as

$$\delta\Pi_1 = \int_A \delta W(\varepsilon_{ij}) \mathrm{d}A = \int_A \sigma_{ij} \delta\varepsilon_{ij} \, \mathrm{d}A, \tag{5.6}$$

$$\delta\Pi_2 = \int_A \Sigma_{ij} \left[\frac{1}{2}\left(\frac{\partial \delta u_i}{\partial X_j} + \frac{\partial \delta u_j}{\partial X_i} \right) - \delta\varepsilon_{ij} \right] \mathrm{d}A, \tag{5.7}$$

$$\delta\Pi_3 = \int_A \delta\Sigma_{ij} \left[\frac{1}{2}\left(\frac{\partial u_i}{\partial X_j} + \frac{\partial u_j}{\partial X_i} \right) - \varepsilon_{ij} \right] \mathrm{d}A, \tag{5.8}$$

$$\delta\Pi_4 = - \int_A F_i \delta u_i \, \mathrm{d}A, \tag{5.9}$$

$$\delta\Pi_5 = - \int_{S_t} T_i \delta u_i \, \mathrm{d}S, \tag{5.10}$$

$$\delta\Pi_6 = \int_{S_u} \delta r_i (\tilde{u}_i - u_i) \mathrm{d}S - \int_{S_u} r_i \delta u_i \, \mathrm{d}S. \tag{5.11}$$

The Fraeijs de Veubeke variational principle expresses that the four proposed fields coincide with those deduced from the theory of elasticity when the first variation of the functional $\Pi = \Pi(u_i, \Sigma_{ij}, \varepsilon_{ij}, r_i)$ vanishes,

$$\delta\Pi = \delta\Pi_1 + \delta\Pi_2 + \delta\Pi_3 + \delta\Pi_4 + \delta\Pi_5 + \delta\Pi_6 = 0. \tag{5.12}$$

This equation constitutes the Fraeijs de Veubeke variational principle.

The first term in equation [5.7] can be integrated in parts:

$$\int_A \Sigma_{ij} \left[\frac{1}{2}\left(\frac{\partial \delta u_i}{\partial X_j} + \frac{\partial \delta u_j}{\partial X_i} \right) \right] \mathrm{d}A = \oint_S N_j \Sigma_{ji} \delta u_i \mathrm{d}S - \int_A \frac{\partial \Sigma_{ji}}{\partial X_j} \delta u_i \, \mathrm{d}A, \tag{5.13}$$

with

$$\oint_S N_j \Sigma_{ji} \delta u_i \, \mathrm{d}S = \int_{S_u} N_j \Sigma_{ji} \delta u_i \, \mathrm{d}S + \int_{S_t} N_j \Sigma_{ji} \delta u_i \, \mathrm{d}S. \tag{5.14}$$

In the case of a linear elastic solid, we have:

$$\sigma_{ij} = \frac{\partial W(\varepsilon_{ij})}{\partial \varepsilon_{ij}}, \tag{5.15}$$

Variation	Equation	Comments
$\delta \varepsilon_{ij}$ in A	$\sigma_{ij} = \Sigma_{ij}$	Proposed stresses are identified as the constitutive stresses in the solid
$\delta \Sigma_{ij}$ in A	$\frac{1}{2}\left(\frac{\partial u_i}{\partial X_j} + \frac{\partial u_j}{\partial X_i}\right) = \varepsilon_{ij}$	Compatibility between the proposed strain and the proposed displacements in the solid
δr_i on S_u	$u_i = \tilde{u}_i$	Compatibility between the proposed displacements and the imposed displacements on the boundary of the solid
δu_i in A	$\frac{\partial \Sigma_{ji}}{\partial X_j} + F_i = 0$	Balance between the proposed stresses and imposed voluminal forces in the solid
δu_i on S_t	$N_j \Sigma_{ji} = T_i$	Balance between the proposed stresses and the surface forces imposed on the boundary of the solid
δu_i on S_u	$N_j \Sigma_{ji} = r_i$	Balance between the proposed stresses and the proposed support reactions on the border of the solid

Table 5.1. *Euler equations deduced from the Fraeijs de Veubeke variational principle*

and

$$W(\varepsilon_{ij}) = \frac{1}{2}C_{ijkl}\varepsilon_{ij}\varepsilon_{kl}, \qquad [5.16]$$

where C_{ijkl} is the classical Hooke tensor.

By substituting these results into equation [5.12], we obtain the Euler equations deduced from the Fraeijs de Veubeke variational principle. These are summarized in Table 5.2.

5.3. Field decomposition

In the constrained natural element method, the domain A contains N nodes, including those on the contour, and the N Voronoi cells are constructed (Figure 5.1).

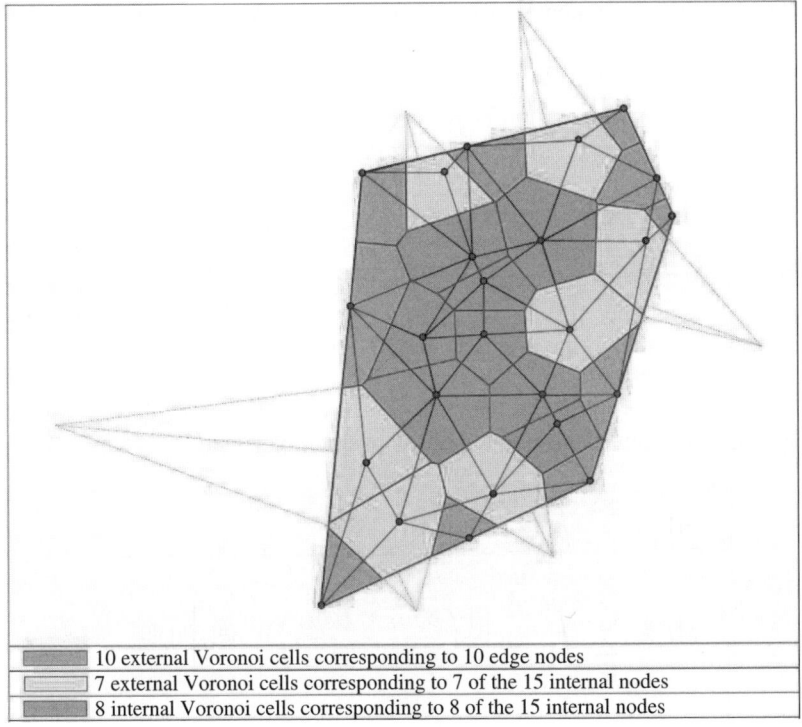

▨	10 external Voronoi cells corresponding to 10 edge nodes
▢	7 external Voronoi cells corresponding to 7 of the 15 internal nodes
▨	8 internal Voronoi cells corresponding to 8 of the 15 internal nodes

Figure 5.1. *Decomposition of domain A in Voronoi cells*

A Voronoi cell that has at least one side belonging to the domain contour is called an exterior Voronoi cell. The corresponding node is not necessarily an interior node. Whereas a Voronoi cell where no side belongs to the contour of the domain is called an interior Voronoi cell. The corresponding node is necessarily an interior node.

The area of the domain is thus:

$$A = \sum_{I=1}^{N} A_I,$$
[5.17]

where A_I is the area of the Voronoi cell I.

We denote C_I the contour of the Voronoi cell I.

In equations [5.10] and [5.11], some integrals are evaluated along the contour of the domain.

This contour is the union of some edges of external Voronoi cells. These sides are noted as S_K, and we have

$$S = \sum_{K=1}^{M} S_K; \qquad S_u = \sum_{K=1}^{M_u} S_K;$$

$$S_t = \sum_{K=1}^{M_t} S_K; \qquad S = S_u \bigcup S_t \Rightarrow M = M_u + M_t,$$

[5.18]

where M is the number of sides that make up the contour, M_u the number of sides on which displacements, $\tilde{u}_i$ are imposed, and M_t the number of sides on which surface tractions T_i are imposed.

By taking account of this domain decomposition, we obtain the following:

$$\delta\Pi_1 = \sum_{I=1}^{N} \int_{A_I} \sigma_{ij} \delta\varepsilon_{ij} \, dA_I,$$

[5.19]

$$\delta\Pi_2 = \sum_{I=1}^{N} \int_{A_I} \Sigma_{ij} \left[\frac{1}{2} \left(\frac{\partial \delta u_i}{\partial X_j} + \frac{\partial \delta u_j}{\partial X_i} \right) - \delta\varepsilon_{ij} \right] dA_I,$$

[5.20]

$$\delta\Pi_3 = \sum_{I=1}^{N} \int_{A_I} \delta\Sigma_{ij} \left[\frac{1}{2} \left(\frac{\partial u_i}{\partial X_j} + \frac{\partial u_j}{\partial X_i} \right) - \varepsilon_{ij} \right] dA_I,$$

[5.21]

$$\delta\Pi_4 = -\sum_{I=1}^{N} \int_{A_I} F_i \delta u_i \, dA_I,$$

[5.22]

$$\delta\Pi_5 = -\sum_{K=1}^{M_t} \int_{S_K} T_i \delta u_i \, dS_K,$$

[5.23]

$$\delta\Pi_6 = \sum_{K=1}^{M_u} \left[\int_{S_K} \delta r_i (\tilde{u}_i - u_i) \, dS_K - \int_{S_K} r_i \delta u_i \, dS_K \right].$$

[5.24]

5.4. Discretization

The following discretization hypotheses are admitted:

– The assumed strain ε_{ij} is constant in each Voronoi cell I:

$$\varepsilon_{ij} = \varepsilon_{ij}^I, \quad I = 1, N.$$

[5.25]

– The assumed stresses Σ_{ij} are constant in each Voronoi cell I:

$$\Sigma_{ij} = \Sigma_{ij}^I, \quad I = 1, N. \tag{5.26}$$

– The assumed support reactions r_i are constant on each side K of the Voronoi cells on which the displacements are imposed:

$$r_i = r_i^K, \quad K = 1, M_u. \tag{5.27}$$

– The assumed displacements u_i are interpolated by the Laplace functions [BEL 97, HIY 99]:

$$u_i = \sum_{J=1}^{N} \Phi_J u_i^J, \tag{5.28}$$

where u_i^J is the assumed displacement of the node J.

Sibson interpolation [SIB 81] can also be used.

Equations [5.15] and [5.25] demonstrate that the constitutive stresses σ_{ij} are constant in each Voronoi cell I:

$$\sigma_{ij} = \sigma_{ij}^I, \quad I = 1, N. \tag{5.29}$$

The variations of the independent variables are

$$\delta\varepsilon_{ij} = \delta\varepsilon_{ij}^I, \quad I = 1, N, \tag{5.30}$$

$$\delta\Sigma_{ij} = \delta\Sigma_{ij}^I, \quad I = 1, N, \tag{5.31}$$

$$\delta r_i = \delta r_i^K, \quad K = 1, M_u, \tag{5.32}$$

$$\delta u_i = \sum_{I=1}^{N} \Phi_I \delta u_i^I. \tag{5.33}$$

By inserting these variations into equations [5.19]–[5.21] and [5.24], after integration by parts gives the following:

$$\delta\Pi_1 = \sum_{I=1}^{N} \sigma_{ij}^I \delta\varepsilon_{ij}^I A_I, \tag{5.34}$$

$$\delta\Pi_2 = \sum_{I=1}^{N} \Sigma_{ij}^I \int_{A_I} \left[\frac{1}{2} \left(\frac{\partial \delta u_i}{\partial X_j} + \frac{\partial \delta u_j}{\partial X_i} \right) \right] dA_I - \sum_{I=1}^{N} \Sigma_{ij}^I \delta\varepsilon_{ij}^I A_I$$

$$= \sum_{I=1}^{N} \Sigma_{ij}^I \oint_{C_I} N_j^I \delta u_i \, dC_I - \sum_{I=1}^{N} \Sigma_{ij}^I \delta\varepsilon_{ij}^I A_I, \tag{5.35}$$

$$\delta\Pi_3 = \sum_{I=1}^{N} \delta\Sigma_{ij}^I \int_{A_I} \left[\frac{1}{2}\left(\frac{\partial u_i}{\partial X_j} + \frac{\partial u_j}{\partial X_i} \right) \right] \mathrm{d}A_I - \sum_{I=1}^{N} \delta\Sigma_{ij}^I \varepsilon_{ij}^I A_I$$

$$= \sum_{I=1}^{N} \delta\Sigma_{ij}^I \oint_{C_I} N_j^I u_i \, \mathrm{d}C_I - \sum_{I=1}^{N} \delta\Sigma_{ij}^I \varepsilon_{ij}^I A_I,$$

[5.36]

$$\delta\Pi_6 = \sum_{K=1}^{M_u} \left[\delta r_i^K \int_{S_K} (\tilde{u}_i - u_i) \mathrm{d}S_K - r_i^K \int_{S_K} \delta u_i \mathrm{d}S_K \right],$$

[5.37]

where N_j^I is the exterior unit normal to the contour of Voronoi cell I.

By substituting these results into the Fraeijs de Veubeke variational principle (equation [5.12]), we obtain

$$\delta\Pi = \delta\Pi_{VA} + \delta\Pi_{VC} + \delta\Pi_{DC} + \delta\Pi_{EF} = 0,$$

[5.38]

where

$$\delta\Pi_{VA} = \sum_{I=1}^{N} \left(\sigma_{ij}^I - \Sigma_{ij}^I \right) \delta\varepsilon_{ij}^I A_I - \sum_{I=1}^{N} \delta\Sigma_{ij}^I \varepsilon_{ij}^I A_I,$$

[5.39]

$$\delta\Pi_{VC} = \sum_{I=1}^{N} \Sigma_{ij}^I \oint_{C_I} N_j^I \delta u_i \mathrm{d}C_I + \sum_{I=1}^{N} \delta\Sigma_{ij}^I \oint_{C_I} N_j^I u_i \mathrm{d}C_I,$$

[5.40]

$$\delta\Pi_{DC} = \sum_{K=1}^{M_u} \left[\delta r_i^K \int_{S_K} (\tilde{u}_i - u_i) \, \mathrm{d}S_K - r_i^K \int_{S_K} \delta u_i \mathrm{d}S_K \right],$$

[5.41]

$$\delta\Pi_{EF} = -\sum_{I=1}^{N} \int_{A_I} F_i \delta u_i \mathrm{d}A_I - \sum_{K=1}^{M_t} \int_{S_K} T_i^K \delta u_i \mathrm{d}S_K.$$

[5.42]

In equations [5.40]–[5.42], the displacements and the virtual displacements are interpolated as specified in equations [5.28] and [5.33], respectively.

By substituting in equation [5.40], we obtain

$$\delta\Pi_{VC} = \sum_{I=1}^{N} \Sigma_{ij}^I \left[\sum_{J=1}^{N} \oint_{C_I} N_j^I \Phi_J \delta u_i^J \mathrm{d}C_I \right] + \sum_{I=1}^{N} \delta\Sigma_{ij}^I \left[\sum_{J=1}^{N} \oint_{C_I} N_j^I \Phi_J u_i^J \mathrm{d}C_I \right].$$

[5.43]

Finally, considering the sides of the Voronoi cells as segments of straight lines, the unit normal to the side S_K is constant along this side and is noted as N_j^K.

Using the discretized displacements of equation [5.28], [5.41] becomes

$$
\delta\Pi_{DC} = \sum_{K=1}^{M_u} \delta r_i^K \left\{ \int_{S_K} \tilde{u}_i dS_K - \sum_{J=1}^{N} u_i^J \int_{S_K} \Phi_J dS_K \right\}
$$
$$
- \sum_{K=1}^{M_u} r_i^K \left\{ \sum_{J=1}^{N} \delta u_i^J \int_{S_K} \Phi_J dS_K \right\}.
$$

[5.44]

In the same way, equation [5.42] becomes

$$
\delta\Pi_{EF} = -\sum_{I=1}^{N}\sum_{J=1}^{N} \delta u_i^J \int_{A_I} F_i \Phi_J \, dA_I - \sum_{K=1}^{M_t}\sum_{J=1}^{N} \delta u_i^J \int_{S_K} T_i \Phi_J \, dS_K.
$$

[5.45]

By combining all these results, the discretized form of the Fraeijs de Veubeke variational principle can be written as

$$
\delta\Pi = \sum_{I=1}^{N} \left(\sigma_{ij}^I - \Sigma_{ij}^I \right) \delta\varepsilon_{ij}^I A_I - \sum_{I=1}^{N} \delta\Sigma_{ij}^I \varepsilon_{ij}^I A_I
$$
$$
+ \sum_{I=1}^{N} \Sigma_{ij}^I \left\{ \sum_{J=1}^{N} \delta u_j^J \oint_{C_I} N_j^I \Phi_J dC_I \right\} + \sum_{I=1}^{N} \delta\Sigma_{ij}^I \left\{ \sum_{J=1}^{N} u_i^J \oint_{C_I} N_j^I \Phi_J dC_I \right\}
$$
$$
\times \sum_{K=1}^{M_u} \delta r_i^K \left\{ \int_{S_K} \tilde{u}_i \, dS_K - \sum_{J=1}^{N} u_i^J \int_{S_K} \Phi_J \, dS_K \right\}
$$
$$
- \sum_{K=1}^{M_u} r_i^K \left\{ \sum_{J=1}^{N} \delta u_i^J \int_{S_K} \Phi_J dS_K \right\} - \sum_{I=1}^{N}\sum_{J=1}^{N} \delta u_i^J \int_{A_I} F_i \Phi_J dA_I
$$
$$
- \sum_{K=1}^{M_t}\sum_{J=1}^{N} \delta u_i^J \int_{S_K} T_i \Phi_J dS_K = 0.
$$

[5.46]

5.5. Discretized equations

We can reorganize the terms of equation [5.46] as follows:

$$\delta\Pi = \sum_{I=1}^{N} \delta\varepsilon_{ij}^{I} \left\{ (\sigma_{ij}^{I} - \Sigma_{ij}^{I}) A_I \right\} + \sum_{I=1}^{N} \delta\Sigma_{ij}^{I} \left\{ -\varepsilon_{ij}^{I} A_I + \sum_{J=1}^{N} u_i^{J} A_j^{IJ} \right\}$$

$$+ \sum_{K=1}^{M_u} \delta r_i^{K} \left\{ \tilde{U}_i^{K} - \sum_{J=1}^{N} u_i^{J} B^{KJ} \right\}$$

$$+ \sum_{J=1}^{N} \delta u_i^{J} \left\{ \sum_{I=1}^{N} \left(\Sigma_{ij}^{I} A_j^{IJ} - \tilde{F}_i^{IJ} \right) - \sum_{K=1}^{M_t} \tilde{T}_i^{KJ} \right\}$$

$$- \sum_{J=1}^{N} \delta u_i^{J} \left\{ \sum_{K=1}^{M_u} r_i^{K} B^{KJ} \right\} = 0,$$

[5.47]

with the notations

$$A_j^{IJ} = \oint_{C_I} N_j^{I} \Phi_J \, dC_I,$$ [5.48]

$$B^{KJ} = \int_{S_K} \Phi_J \, dS_K,$$ [5.49]

$$\tilde{U}_i^{K} = \int_{S_K} \tilde{u}_i \, dS_K,$$ [5.50]

$$\tilde{F}_i^{IJ} = \int_{A_I} F_i \Phi_J \, dA_I,$$ [5.51]

$$\tilde{T}_i^{KJ} = \int_{S_K} T_i \Phi_J \, dS_K.$$ [5.52]

Equation [5.48] implies an integration on the contour C_I of the Voronoi cell I.

Equations [5.49] and [5.50] imply an integration on the side S_K (pertaining to the field contour) of an exterior Voronoi cell.

Thus, we can write the Euler equations deduced from the discretized variational principle as follows:

– In each Voronoi cell I:

$$\sigma_{ij}^{I} = \Sigma_{ij}^{I}, \quad I = 1, N.$$ [5.53]

These equations identify the assumed stresses Σ_{ij}^{I} with the constitutive stresses σ_{ij}^{I} deduced from the assumed strains ε_{ij}^{I} in each Voronoi cell.

– In each Voronoi cell I:

$$\varepsilon_{ij}^I A_I = \sum_{J=1}^{N} u_i^J A_j^{IJ}, \quad I = 1, N. \qquad [5.54]$$

These are compatibility equations connecting the assumed strains ε_{ij}^I in each Voronoi cell I with assumed nodal displacements u_i^J.

– On the sides K of the Voronoi cells subjected to imposed displacements:

$$\sum_{J=1}^{N} u_i^J B^{KJ} = \tilde{U}_i^K, \quad K = 1, M_u. \qquad [5.55]$$

These are also compatibility equations between the $\tilde{u}_i$ displacements imposed on the S_u part contour domain and the assumed nodal displacements u_i^J.

– In each Voronoi cell J:

$$\sum_{I=1}^{N} \left(\Sigma_{ij}^I A_j^{IJ} - \tilde{F}_i^{IJ} \right) - \sum_{K=1}^{M_t} \tilde{T}_i^{KJ} - \sum_{K=1}^{M_u} r_i^K B^{KJ} = 0, \quad J = 1, N. \qquad [5.56]$$

These are equations expressing the balance of the body F_i, surface tractions T_i, and the assumed support reactions r_i with the assumed stresses Σ_{ij}^I in the Voronoi cells.

We note that in these results, the only term that requires an integration on the area of the Voronoi cells is $\tilde{F}_i^{IJ} = \int_{A_I} F_i \Phi_J \, dA_I$.

Consequently, in the absence of body forces, only the integrals along the sides of the Voronoi cells remain.

We can then use a conventional Gaussian integration scheme. In this case, there is no reason to resort to specific schemes of integration [ATL 99, ATL 00, CUE 03] on the area of the domain. We also note that this formulation avoids the calculation of the derivatives of the Laplace or Sibson interpolation function.

Thus, based on the Fraeijs de Veubeke variational principle, we recognize the same advantages as with stabilized integration [CHE 01, YOO 04].

Finally, the mixed approach developed here shows that it is possible to impose $\tilde{u}_i$ displacements on any side of K composing the discretized domain contour. They can be an any function $\tilde{u}_i = \tilde{u}_i(s)$ of an abscissa s along the contour. From equations [5.50] and [5.55], we can note that they are satisfied in a weighted average sense.

5.6. Matrix solution for linear elastic problems

The discretized equations [5.53]–[5.56] carry a large number of unknowns, which seems to be an important disadvantage of the mixed approach developed earlier. We can show that this is not the case, however, if we examine the solution to the equation in great detail. To this end, we introduce the matrix notations as defined in Table 5.6.

Therefore, we successively obtain

$$\sigma_{ij}^I = \Sigma_{ij}^I \;\Rightarrow\; \{\sigma\}^I = \{\Sigma\}^I, \tag{5.57}$$

$$\Sigma_{ij}^I A_j^{IJ} \;\Rightarrow\; [A]^{IJ}\{\Sigma\}^I; \quad \tilde{F}^{IJ} \;\Rightarrow\; \{\tilde{F}\}^{IJ}; \quad \tilde{T}^{IJ} \;\Rightarrow\; \{\tilde{T}\}^{IJ}, \tag{5.58}$$

$$\sum_{I=1}^N \Sigma_{ij}^I A_j^{IJ} - \sum_{K=1}^{M_u} r_i^K B^{KJ} = \sum_{I=1}^N \tilde{F}_i^{IJ} + \sum_{K=1}^{M_t} \tilde{T}_i^{KJ}$$

$$\Rightarrow \sum_{I=1}^N [A]^{IJ}\{\Sigma\}^I - \sum_{K=1}^{M_u} B^{KJ}\{r\}^K = \{\tilde{F}\}^J + \{\tilde{T}\}^J. \tag{5.59}$$

The term $\sum_{I=1}^N [A]^{IJ}\{\Sigma\}^I - \sum_{K=1}^{M_u} B^{KJ}\{r\}^K$ can be interpreted as an energetically equivalent "interior" nodal force applied to the node J.

This force is made up of the contribution $[A]^{IJ}\{\Sigma\}^I$ of the assumed stresses Σ_{ij}^I present in the Voronoi cells I and the contribution $B^{KJ}\{r\}^K$ of the assumed support reactions t_i^K present on the sides K of the contour on which the displacements are imposed.

The term $\{\tilde{F}\}^J + \{\tilde{T}\}^J$ can be interpreted as an energetically equivalent "exterior" nodal force applied to the node J.

It is the sum:

– of the contributions $\{\tilde{F}\}^{IJ}$ of the body forces F_i applied to the Voronoi cells I;

– of the contributions $\{\tilde{T}\}^{IJ}$ of the surface tractions T_i applied to the part S_t of the domain contour.

Let us now consider equation [5.53]. In matrix notation, this equation is written as

$$A_I\{\varepsilon\}^I = \sum_{J=1}^N [A]^{IJ,T}\{u\}^J \quad I = 1, N, \tag{5.60}$$

where $[A]^{IJ,T}$ is the transpose of $[A]^{IJ}$.

It is noted that in $\{\varepsilon\}^I$, the third component is $2\varepsilon_{12}^I$.

Notations	Comments
$\{\varepsilon\}^I = \left\{\begin{array}{c} \varepsilon_{11}^I \\ \varepsilon_{22}^I \\ 2\varepsilon_{12}^I \end{array}\right\}$	Assumed strains in the Voronoi cell I
$\{\sigma\}^I = \left\{\begin{array}{c} \sigma_{11}^I \\ \sigma_{22}^I \\ \sigma_{12}^I \end{array}\right\}$	Assumed in the Voronoi cell I
$\{u\}^I = \left\{\begin{array}{c} u_1^I \\ u_2^I \end{array}\right\}$	Assumed nodal displacements of node I
$\{t\}^K = \left\{\begin{array}{c} t_1^K \\ t_2^K \end{array}\right\}$	Assumed support reactions on the side K subjected to imposed displacements
$A_I; C_I$	Area and contour of the Voronoi cell I
S_K	Length of the side K of the Voronoi cell
Φ_J	Interpolation function associated with node J
$\tilde{F}_i^{IJ} = \int_{A_I} F_i \Phi_J \mathrm{d}A_I;$ $\{\tilde{F}\}^{IJ} = \left\{\begin{array}{c} \tilde{F}_1^{IJ} \\ \tilde{F}_2^{IJ} \end{array}\right\}; \{\tilde{F}\}^J = \sum_{I=1}^N \{\tilde{F}\}^{IJ}$	$\{\tilde{F}\}^J$ is the nodal force on node J, which is energetically equivalent to the body forces F_i applied to the solid
$\tilde{T}_i^{KJ} = \int_{S_K} T_i \Phi_J \mathrm{d}S_K;$ $\{\tilde{T}\}^{KJ} = \left\{\begin{array}{c} \tilde{T}_1^{KJ} \\ \tilde{T}_2^{KJ} \end{array}\right\};$ $\{\tilde{T}\}^J = \sum_{K=1}^{M_t} \{\tilde{T}\}^{KJ}$	$\{\tilde{T}\}^J$ is the nodal force on the node J, which is energetically equivalent to the surface tractions T_i applied to the contour of solid
$\tilde{U}_i^K = \int_{S_K} \tilde{u}_i \, \mathrm{d}S_K;$ $\{\tilde{U}\}^K = \left\{\begin{array}{c} \tilde{U}_2^K \\ \tilde{U}_1^K \end{array}\right\}$	$\{\tilde{U}\}^K$ is a generalized displacement equivalent to the imposed $\tilde{u}_i$ displacements on the side K
$B^{KJ} = \int_{S_K} \Phi_J \mathrm{d}S_K$	Interpolation function integrated on the side K of a Voronoi cell
$A_j^{IJ} = \oint_{C_I} N_j^I \Phi_J \mathrm{d}C_I;$ $[A]^{IJ} = \left[\begin{array}{ccc} A_1^{IJ} & 0 & A_2^{IJ} \\ 0 & A_2^{IJ} & A_1^{IJ} \end{array}\right]$	N_j^I is the exterior unit normal to the contour of the Voronoi cell I

Table 5.2. *Matrix notation*

The equation of compatibility [5.60] defines the assumed strain $\{\varepsilon\}^I$ in the Voronoi cell I as the sum of the contributions $[A]^{IJ,T}\{u\}^J$ of all the nodes J.

Finally, the equation of compatibility [5.55] relating to the sides K subjected to imposed displacements becomes

$$\sum_{J=1}^{N} B^{KJ}\{u\}^{J} = \{\tilde{U}\}^{K} \quad K = 1, M_u.$$ [5.61]

Table 5.6 list of the equations in matrix notation.

Equations	Comments	
$\{\sigma\}^{I} = \{\Sigma\}^{I}, I = 1, N$	Identification of the assumed stresses with the constitutive stresses in cell I	[5.62]
$\sum_{I=1}^{N}[A]^{IJ}\{\Sigma\}^{I} - \sum_{K=1}^{M_u} B^{KJ}\{r\}^{K}$ $= \{\tilde{F}\}^{J} + \{\tilde{T}\}^{J}, J = 1, N$	Equilibrium equation in cell J	[5.63]
$A_I\{\varepsilon\}^{I} = \sum_{J=1}^{N}[A]^{IJ,T}\{u\}^{J}, I = 1, N$	Compatibility equation in cell I	[5.64]
$\sum_{J=1}^{N} B^{KJ}\{u\}^{J} = \{\tilde{U}\}^{K}, K = 1, M_u$	Compatibility equation on the side K subjected to imposed displacement	[5.65]

Table 5.3. *Discretized equations in matrix notation*

For a linear elastic material, Hooke discretized law is written in the Voronoi cell J as

$$\{\sigma\}^{J} = [C]^{J}\{\varepsilon\}^{J},$$ [5.66]

where $[C]^{J}$ is the Hooke matrix for material constituting the Voronoi J.

By substituting equation [5.64] in equation [5.66], we obtain

$$\{\sigma\}^{J} = [C]^{J}\{\varepsilon\}^{J} = [C^{*}]^{J} \sum_{J=1}^{N}[A]^{IJ,T}\{u\}^{J},$$ [5.67]

with

$$[C^*]^J = \frac{1}{A^J}[C]^J.$$ [5.68]

Consequently, using equation [5.62], the first term of equation [5.63] becomes

$$\sum_{I=1}^{N}[A]^{IJ}\{\Sigma\}^I = \sum_{L=1}^{N}\left(\sum_{I=1}^{N}[A]^{IJ}[C^*]^I[A]^{IL,T}\right)\{u\}^L = \sum_{L=1}^{N}[M]^{JL}\{u\}^L,$$ [5.69]

with

$$[M]^{JL} = \sum_{I=1}^{N}[A]^{IJ}[C^*]^I[A]^{IL,T}.$$ [5.70]

By substituting the above value in equation [5.62], we obtain

$$\sum_{L=1}^{N}[M]^{JL}\{u\}^L = \sum_{K=1}^{M_u}B^{KJ}\{r\}^K + \{\tilde{F}\}^J + \{\tilde{T}\}.$$ [5.71]

Equations [5.71] and [5.65] constitute a system of equations of the form:

$$\begin{bmatrix} [M] & -[B] \\ -[B]^T & [0] \end{bmatrix}\begin{Bmatrix} \{q\} \\ \{r\} \end{Bmatrix} = \begin{Bmatrix} \{\tilde{Q}\} \\ -\{\tilde{U}\} \end{Bmatrix}$$ [5.72]

with

$$\{q\} = \begin{Bmatrix} \{u\}^1 \\ \{u\}^2 \\ . \\ . \\ . \\ \{u\}^N \end{Bmatrix}; \{\tilde{F}\} = \begin{Bmatrix} \{\tilde{F}\}^1 \\ \{\tilde{F}\}^2 \\ . \\ . \\ \{\tilde{F}\}^N \end{Bmatrix}; \{\tilde{T}\} = \begin{Bmatrix} \{\tilde{T}\}^1 \\ \{\tilde{T}\}^2 \\ . \\ . \\ \{\tilde{T}\}^N \end{Bmatrix};$$ [5.73]

$$\{\tilde{Q}\} = \{\tilde{F}\} + \{\tilde{T}\}$$

$$[M] = \begin{bmatrix} [M]^{11} & [M]^{12} & . & . & [M]^{1N} \\ [M]^{21} & [M]^{22} & . & . & [M]^{2N} \\ . & . & . & . & . \\ . & . & . & . & . \\ [M]^{N1} & [M]^{N2} & . & . & [M]^{NN} \end{bmatrix}$$ [5.74]

We can easily verify that the matrix $[M]$ is symmetrical.

The equation [5.65] which, in matrix form, is written as $[B]^T\{q\} = \{\tilde{U}\}$ constitutes a group of linear constraints linking the nodal displacements $\{q\}$. In particular, if displacements $\tilde{u}_i = 0$ are imposed on segment AB connecting two nodes A and B of the domain contour, we can easily show that equation [5.65] gives $u_i^A = 0$ and $u_i^B = 0$.

Displacements u_i^A and u_i^B can then be directly eliminated from $\{q\}$. This reasoning extends easily if null displacements are imposed on any number of segments of the contour.

Thus, despite the large number of initial unknowns resulting from the discretization (equations [5.25]–[5.28]), we finally come back to a classical system of equations of the form $[M]\{q\} = \{\tilde{Q}\}$, which has the same characteristics as in the approach based on the virtual work principle.

5.7. Numerical integration

To numerically evaluate all the integrals of the type $\int_{S_K}(\bullet)dS_K$ on side K of a Voronoi cell, the Gauss method can be used.

By defining a local coordinate $-1 \leq \xi \leq +1$ along the side S_K, such an integral takes the form:

$$\int_{-1}^{+1} f(\xi)d\xi = \sum_{IP=1}^{NP} f(\xi_{IP})W(IP),$$

where IP is the integration point, NP the number of integration points, ξ_{IP} the coordinate of point IP, and $W(IP)$ the integration weight of point IP.

The precision of this diagram was tested for NP varying from 1 to 10.

To this end, we have taken advantage of the fact that the expression $A_j^{IJ} = \oint_{C_I} N_j^I \Phi_J\, dC_I$ can be analytically calculated for a regular distribution of nodes ("squared" distribution), which is illustrated in Figure 5.2.

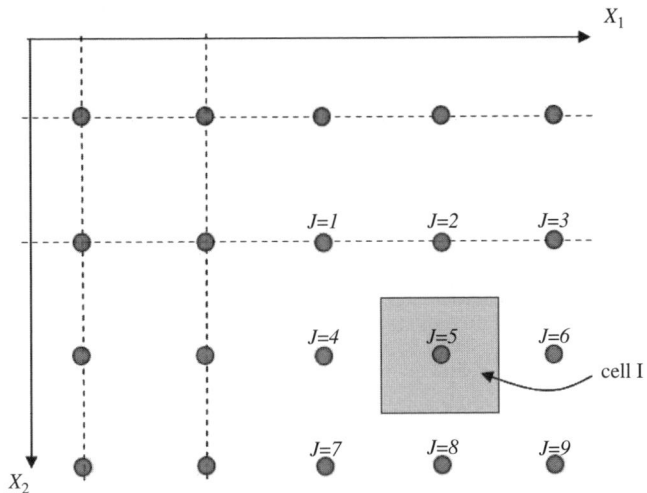

Figure 5.2. *"Squared" distribution of nodes*

Tables 5.4 and 5.5 give two examples of the results obtained for the relative error on the calculation of terms A_1^{IJ} and A_2^{IJ} for one of the nodes ($J = 1, 9$) identified in Figure 5.2. We know that, for reasons of symmetry, the errors are the same for any selected value of $J = 1, 9$ while for the other nodes, $\Phi_J = 0$.

Points of integration	NP= 1	NP= 2	NP= 3	NP= 4	NP= 5
Relative error on $A_1^{IJ}(\%)$	30.13	0.77	4.52	2.62	0.75
Relative error on $A_2^{IJ}(\%)$	12.82	0.32	1.85	1.08	0.31

Table 5.4. *Precision of numerical integration (NP= 1 with NP= 5)*

Points of integration	NP= 6	NP= 7	NP= 8	NP= 9	NP= 10
Relative error on $A_1^{IJ}(\%)$	0.99	1.72	0.59	0.24	0.57
Relative error on $A_2^{IJ}(\%)$	0.41	0.71	0.24	0.10	0.24

Table 5.5. *Precision of numerical integration (NP= 6 with NP= 10)*

We noted that the precision does not increase in a monotonic way with the number of Gauss points.

By retaining NP=2, precision can be accepted at the least cost.

5.8. Linear elastic patch tests

A series of patch tests in simple tension and pure shear makes it possible to validate the mixed approach. Figure 5.3 shows the studied fields and their discretizations in Voronoi cells. Note that the nodes are not represented.

Number of nodes	Voronoi cells
Discretization 1 4 nodes	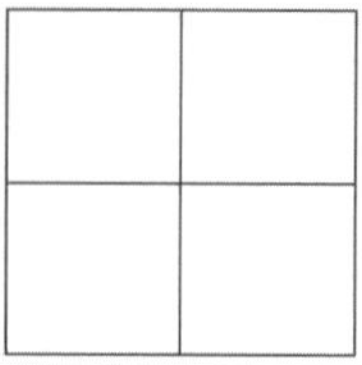
Discretization 2 45 nodes	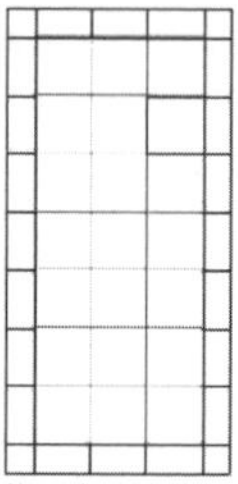
Discretization 3 38 nodes	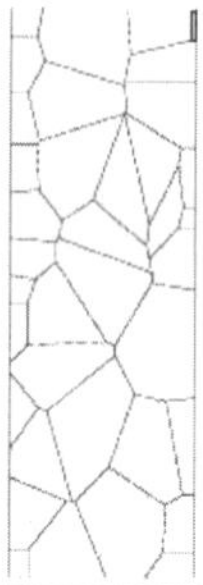

Figure 5.3. *Studied fields and Voronoi cells*

The results are shown in Tables 5.6 and 5.7 for two values of Poisson's ratio $\nu = 0$ and $\nu = 0.3$. They are expressed using the following variables:

$$\text{Average} = \frac{\sum\limits_{K=1}^{N} \sigma^K}{N}; \qquad L2\sigma = \frac{\sum\limits_{K=1}^{N} A_K \sqrt{\left(\sigma_{ij}^K - \sigma_{ij}^{\text{exact}}\right)\left(\sigma_{ij}^K - \sigma_{ij}^{\text{exact}}\right)}}{\sum\limits_{K=1}^{N} A_K \sqrt{\sigma_{ij}^{\text{exact}} \sigma_{ij}^{\text{exact}}}}$$

Field discretization	Loading (stress applied)	Number of nodes	σ_{11}	σ_{22}	σ_{12}	$L2\sigma$
			Average (MPa)	Average (MPa)	Average (MPa)	
Discretization 1 regular cells	Tension (1,000 MPa)	4	1,000	−4.42E-12	1.11E-11	2.87E-16
Discretization 2 regular cells	Tension (1,000 MPa)	45	2.94E-13	1,000	−3.67E-12	5.11E-17
Discretization 3 irregular cells	Tension (1,000 MPa)	38	−7.44E-12	1,000	2.16E-11	9.97E-16
Discretization 1 regular cells	Shear (1,000 MPa)	4	1.88E-12	−3.99E-17	1,000	3.77E-16
Discretization 3 irregular cells	Shear (1,000 MPa)	38	8.85E-12	7.13E-12	1,000	7.43E-16

Table 5.6. *Patch tests results for* $E = 10^8$ *MPa and* $\nu = 0$

Field discretization	Loading (stress applied)	Number of nodes	σ_{11} Average (MPa)	σ_{22} Average (MPa)	σ_{12} Average (MPa)	$L2\sigma$
Discretization 1 regular cells	Tension (1,000 MPa)	4	1,000	−4.42E-12	1.11E-10	2.87E-16
Discretization 2 regular cells	Tension (1,000 MPa)	45	2.94E-12	1,000	−3.67E-12	5.11E-17
Discretization 3 irregular cells	Tension (1,000 MPa)	38	−1.44E-12	1,000	5.16E-11	3.10E-16
Discretization 1 regular cells	Shear (1,000 MPa)	4	1.92E-12	−1.74E-13	−1,000	1.77E-16
Discretization 3 irregular cells	Shear (1,000 MPa)	38	1.37E-12	−4.64E-12	−1,000	7.43E-16

Table 5.7. *Patch test results for* $E = 10^8$ *MPa and* $\nu = 0.3$

We note that the patch tests are satisfactory: the error is of the order of machine precision (double precision).

These results were calculated with two Gaussian points on each side of the Voronoi cells (NP= 2). The results with NP= 9 are the same, except for the machine precision.

For discretization 3 (38 nodes, irregular cells), the patch tests in tension were undertaken for a quasi-incompressible material with $\nu = 0.49$, $\nu = 0.499$, and $\nu = 0.4999$.

The results are summarized in Table 5.8, which reveals a loss of precision, as in $\nu \to 0.5$.

This problem is related to the deterioration of the conditioning of the matrix $[M]$ [WIL 65, FRI 73].

If a small error $[M_e]$ exists on $[M]$ because of the machine precision, the solution of the equation system $[M]\{q\} = \{\tilde{Q}\}$ is modified and becomes $\{q\} + \{q_e\}$.

According to [WIL 65], the error can be evaluated by

$$\frac{\|q_e\|}{\|q\|} \le C \frac{\|M_e\|}{\|M\|} \bigg/ \left[1 - C \frac{\|M_e\|}{\|M\|}\right]$$

where $\|\ \|$ represents the L2 norm.

If $\|M_e\| \ll \|M\|$, this result can be approached by

$$\frac{\|q_e\|}{\|q\|} \le C \frac{\|M_e\|}{\|M\|}$$

where C is the conditioning number of $[M]$.

ν	$L2\sigma$
0.3	3.10E-16
0.49	5.14E-15
0.499	1.95E-14
0.4999	4.70E-13

Table 5.8. *Patch tests in tension for a quasi-incompressible material; 38 nodes, irregular cells*

For a three-dimensional solid discretized in tetrahedral finite elements of size h, it is established in [FRI 73] that

$$C = c\frac{h^{-2}}{(1+\nu)(1-2\nu)},$$

where c is a constant.

Although this result was not established within the natural element method framework, a simple calculation shows that when ν goes from 0.49 to 0.499 or 0.4999, the error on the solution is multiplied by 10 or 100 or 1,000. This can explain the results found in Table 5.8.

We can, therefore, consider that the mixed approach of the natural elements proposed here makes it possible to pass the patch test for a quasi-incompressible material.

5.9. Application 1: pure bending of a linear elastic beam

The convergence of the mixed approach is illustrated by a simple case the analytical solution of which is known: the pure bending of a beam with a rectangular-cross-section.

The boundary conditions and loading are shown in Figure 5.4.

Horizontal displacements on the $X = 0$ side are imposed null. To prevent any rigid body motion, a small appendix is added in the middle of this side and is fixed in the two directions.

An example of discretization in 222 irregular Voronoi cells is also shown in this figure.

The strain energy stored in the beam has the following theoretical value:

$$W_{I,th} = \frac{1}{2}\int_A \sigma_{ij}\varepsilon_{ij}\,\mathrm{d}A = \frac{1}{2}\frac{M^2 L(1-\nu^2)}{EI}$$

with

$$M = \int_A \sigma Y\,\mathrm{d}A = \frac{\sigma_{\max}}{h}I; \qquad I = \int_A y^2\,\mathrm{d}A = \frac{2h^3}{3}; \qquad \sigma_{xx} = \frac{M}{I}Y.$$

Furthermore, its value deduced from the numerical calculation is expressed by:

$$W_{I,\mathrm{num}} = \sum_{K=1}^{N}\frac{1}{2}\{\sigma^K\}^T\{\varepsilon^K\}A_K.$$

For $E = 10^8$ MPa, $\nu = 0.3$, and the chosen values of $\sigma_{\max}, h, L$ (Figure 5.4), we obtain $W_{I,th} = 75.8$ J.

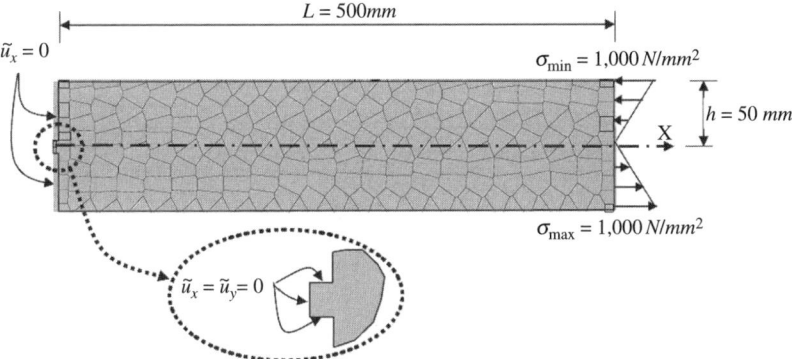

Figure 5.4. *Limit conditions and loading for pure elastic bending of a beam*

To study the influence of the number of nodes and of their distribution on the quality of the solution, the following procedure is developed:

– Create a regular distribution of $(n_X + 1)$ nodes in the direction X and $(n_Y + 1)$ nodes in the direction Y; the spacing of the nodes in directions X and Y is thus: $\Delta_X = L/n_X$ and $\Delta_Y = 2h/n_Y$

– Move the interior nodes by the quantities: $d_X = \frac{\text{random}}{5} d_r \Delta_X$ and $d_Y = \frac{\text{random}}{5} d_r \Delta_Y$, where $-0.5 \leq d_r \leq +0.5$ is a random number with uniform distribution and $0 \leq \text{random} \leq 5$ is a value that can be adjusted to accentuate the irregularity of the nodal distribution; the nodes that leave the domain following these displacements are eliminated.

The results of various calculations are summarized in Table 5.9 and Figures 5.5 and 5.6.

Figure 5.5 shows that convergence toward the theoretical solution is fully obtained, but it is influenced by the regularity of the mesh: convergence is better when the mesh is regular.

For the case: random $= 0.1$, $n_x = 20$, $ny=8$, Figure 5.6 shows the stresses σ_{11} obtained.

To study the effect of quasi-incompressibility, calculations were also carried out for the following values of Poisson's ratio: $\nu = 0.49, 0.499$, and 0.4999. Finally, the convergence in displacements was also evaluated using the following norm:

$$L2u = \frac{\sum_{i=1}^{N} \sqrt{(u_i^{\text{analy}} - u_i^{\text{num}}) \cdot (u_i^{\text{analy}} - u_i^{\text{num}})}}{LN},$$

nX	nY	Number of nodes	$W_{I,\text{num}}\ (J)$
Random = 0.1			
10	4	76	76.9
15	6	140	76.5
20	8	244	76.3
20	16	399	76.0
Random = 3			
10	4	78	80.7
15	6	140	78.9
20	8	222	77.8
20	16	398	76.7

Table 5.9. *Convergence of the strain energy (values)*

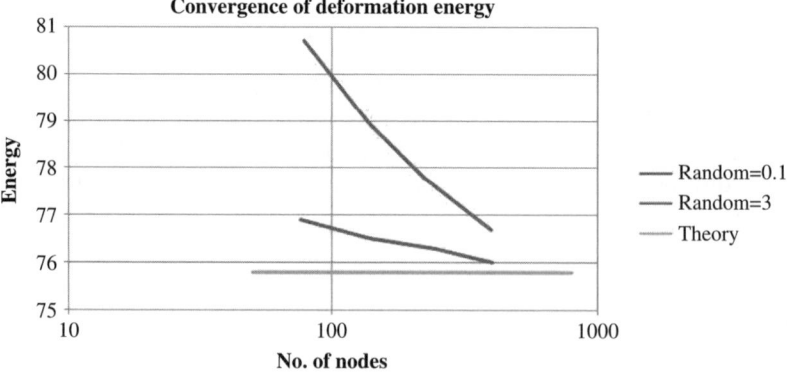

Figure 5.5. *Convergence of the strain energy*

where u_i^{analy} are the nodal displacements deduced from the analytical solution and u_i^{num} are the nodal displacements obtained by the mixed natural elements method. N is the number of nodes.

Figures 5.7 and 5.8 illustrate the convergence of the displacements and of the strain energy for various values of Poisson's ratio. Although there is no formal proof, these results in conjunction with the patch tests tend to show that incompressibility locking is avoided by the formulation developed in this chapter.

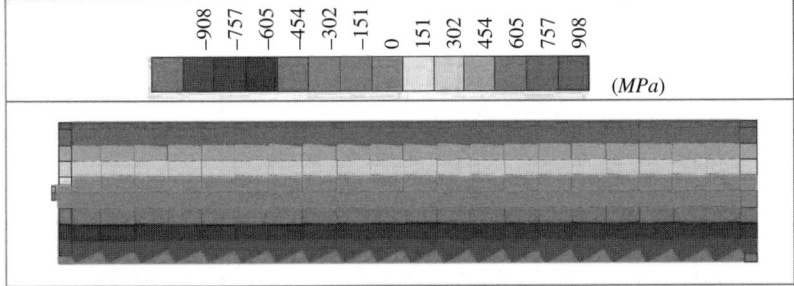

Figure 5.6. *Longitudinal stresses* σ_{11}

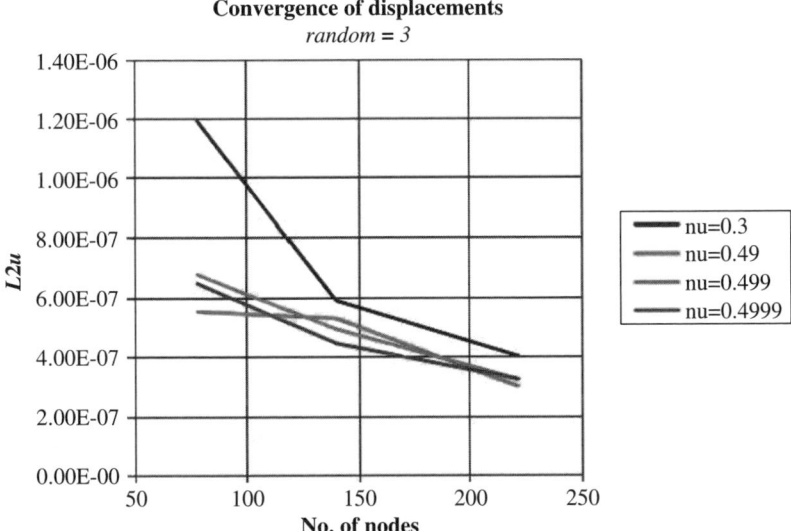

Figure 5.7. *Convergence of displacements for various values of Poisson's ratio*

5.10. Application 2: square domain with circular hole

Let us consider the domain in Figure 5.9. This domain is in a state of strain plane and is calculated with various numbers of nodes and various numbers of Gauss integration points on each side of the Voronoi cells.

We examined the convergence of the results for the strain energy (Table 5.10 and Figure 5.10) and for the stress concentration coefficient (Table 5.11 and Figure 5.11) calculated by $k = \frac{\sigma_{11,\text{max}}}{\sigma_{\text{net}}}$ with $\sigma_{\text{net}} = \frac{\text{total force in direction } X_1}{\text{net section in } X_1=0} = \frac{1,000 \times 600}{600-200} = 1,500 \, MPa$.

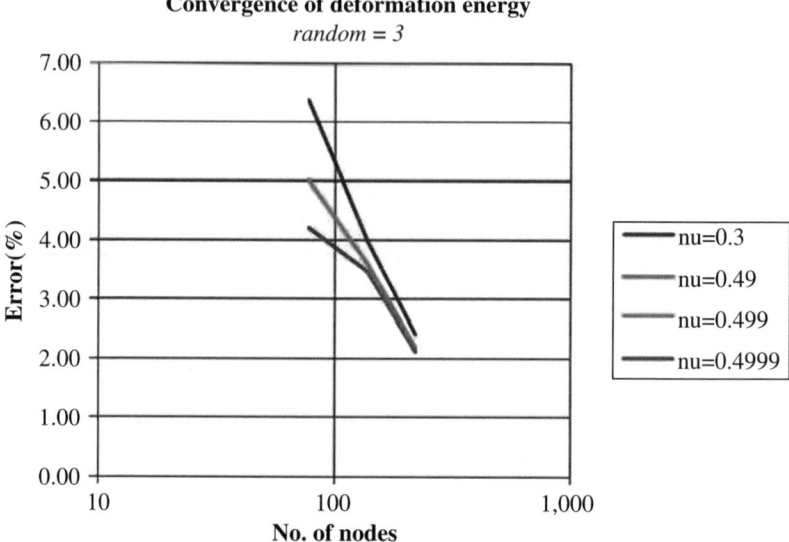

Figure 5.8. *Convergence of the strain energy for various values of Poisson's ratio*

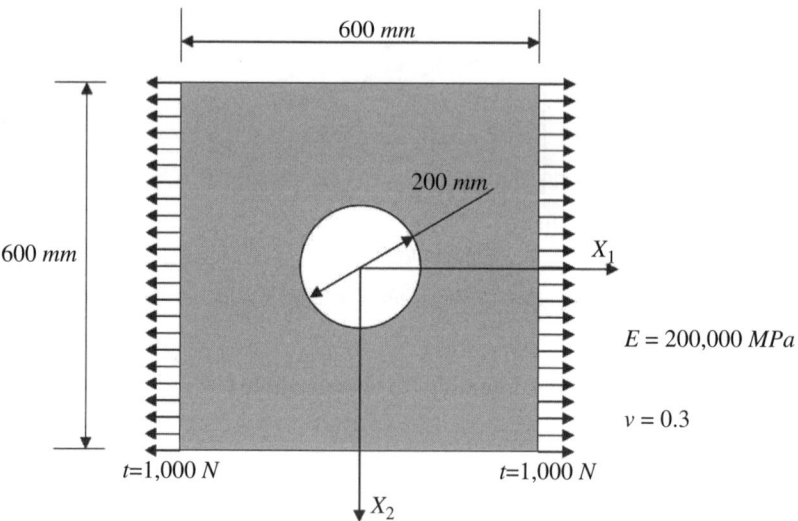

Figure 5.9. *Square domain with circular hole*

In Figure 5.10, we have also computed the convergence curve by the finite elements method. We used isoparametric quadrilateral elements of the first degree integrated with four Gaussian points.

Number of nodes	1 integration point	2 integration points	10 integration points
36	290614	285524	286084
121	284496	282707	282874
441	281741	281070	281146
1,681	280715	280517	280542

Table 5.10. *Convergence of the strain energy (J)*

Number of nodes	1 integration point	2 integration points	10 integration points
36	2.4425	2.3849	2.3731
121	2.5436	2.4994	2.5011
441	2.6068	2.5942	2.5934
1,681	2.6624	2.6638	2.6621

Table 5.11. *Convergence of the stress concentration coefficient*

Figure 5.12 shows the stress maps:

– for σ_{11}: minimum value $= -400\,MPa$, maximum value $= 4,000\,MPa$;

– for σ_{22}: minimum value $= -1,800\,MPa$, maximum value $= 1,000\,MPa$;

– for σ_{12}: minimum value $= -1,200\,MPa$, maximum value $= 400\,MPa$.

It is interesting to note that from Figure 5.10 the finite elements method when used with an equal number of nodes is further removed from the exact solution than with the natural elements method.

This confirms the fact that the natural elements method is less sensitive to the nodal position (with the distortions of the mesh) than the finite elements method because the distortions are all more important when the number of nodes is small.

5.11. Mixed approach to nonlinear problems

The formulation developed here can be expanded to the materially nonlinear problems, which means for problems in which the displacements and strains that

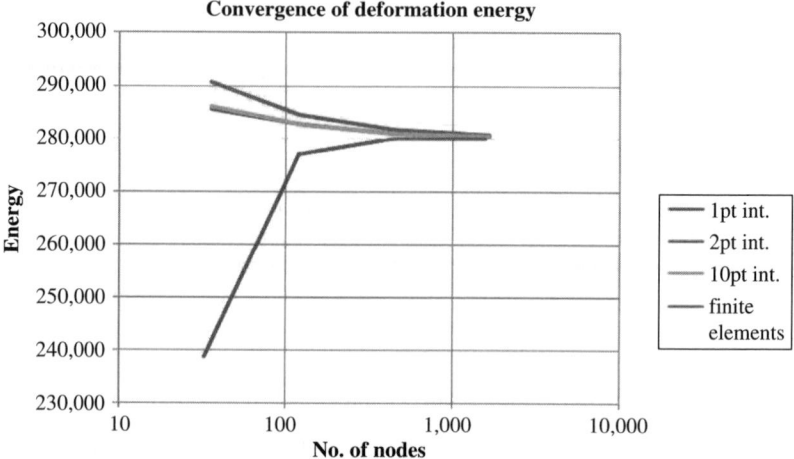

Figure 5.10. *Convergence of the strain energy*

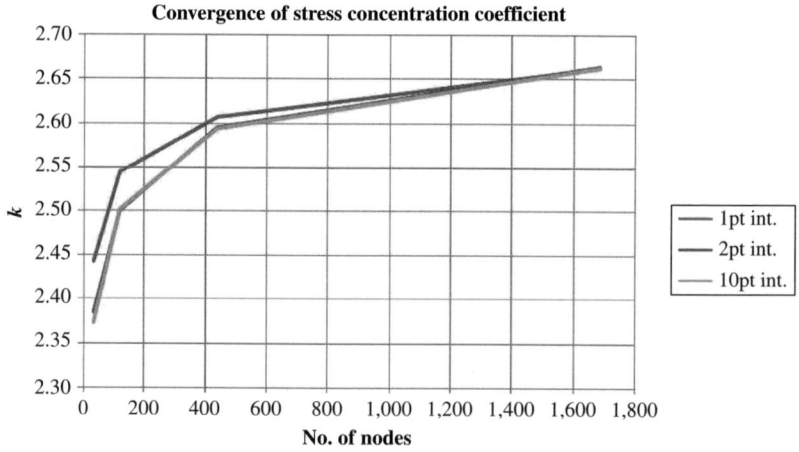

Figure 5.11. *Stress concentration coefficient convergence*

remain weak, but the behavior of the material is not expressed by Hooke's law. For example, this behavior can be elastoplastic or elastoviscoplastic.

In this case, the starting point is the following variational equation which constitutes an obvious extension of the Fraeijs de Veubeke variational principle.

$$\delta\Pi_1 = \int_A \sigma_{ij}\delta\dot{\varepsilon}_{ij}\,dA, \qquad\qquad [5.75]$$

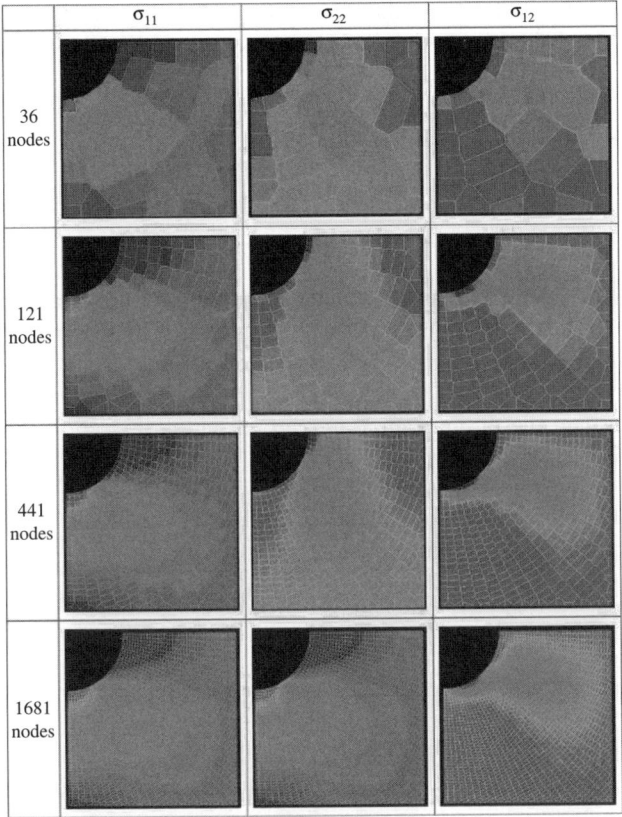

Figure 5.12. *Compared stress maps*

$$\delta\Pi_2 = \int_A \Sigma_{ij} \left[\frac{1}{2}\left(\frac{\partial \delta \dot{u}_i}{\partial X_j} + \frac{\partial \delta \dot{u}_j}{\partial X_i} \right) - \delta\dot{\varepsilon}_{ij} \right] \mathrm{d}A, \qquad [5.76]$$

$$\delta\Pi_3 = \int_A \delta\Sigma_{ij} \left[\frac{1}{2}\left(\frac{\partial \dot{u}_i}{\partial X_j} + \frac{\partial \dot{u}_i}{\partial X_j} \right) - \dot{\varepsilon}_{ij} \right] \mathrm{d}A, \qquad [5.77]$$

$$\delta\Pi_4 = -\int_A F_i \delta \dot{u}_i \, \mathrm{d}A, \qquad [5.78]$$

$$\delta\Pi_5 = -\oint_{S_t} T_i \delta \dot{u}_i \, \mathrm{d}S, \qquad [5.79]$$

$$\delta\Pi_6 = \int_{S_u} \delta r_i \left(\dot{\bar{u}}_i - \dot{u}_i \right) \mathrm{d}S - \int_{S_u} r_i \delta \dot{u}_i \, \mathrm{d}S, \qquad [5.80]$$

$$\delta\Pi\left(\dot{u}_i, \dot{\varepsilon}_{ij}, \Sigma_{ij}, r_i \right) = \delta\Pi_1 + \delta\Pi_2 + \delta\Pi_3 + \delta\Pi_4 + \delta\Pi_5 + \delta\Pi_6 = 0, \qquad [5.81]$$

with $\dot{u}_i$ as the velocity field, $\dot{\varepsilon}_{ij}$ the strain rate field, Σ_{ij} the stress field, and r_i the support reactions field. In equation [5.75], σ_{ij} are the constitutive stresses.

For inelastic materials, they are generally obtained by time integration of a system of differential equations of the following type:

$$\dot{\sigma}_{ij} = f_{ij}\left(\sigma_{ij}, q_{ij}, \dot{\varepsilon}_{ij}\right), \qquad [5.82]$$

$$\dot{q}_{ij} = h_{ij}\left(\sigma_{ij}, q_{ij}\right), \qquad [5.83]$$

where q_{ij} are the internal variables, which can be scalar or tensorial according to the case. The Euler equations deduced from [5.81] are summarized in Table 5.11. They are very similar to those in Table 5.2. The difference comes from the replacement of displacements u_i by velocities $\dot{u}_i$ and strains ε_{ij} by the strain rates $\dot{\varepsilon}_{ij}$. The domain decomposition and the discretization are carried out as in sections 5.3 and 5.4 by means of replacing u_i with $\dot{u}_i$ and ε_{ij} with $\dot{\varepsilon}_{ij}$.

At this stage, it is appropriate to note an important point. In the discretization, we have chosen a constant strain rate field in each Voronoi cell.

$$\dot{\varepsilon}_{ij} = \dot{\varepsilon}_{ij}^I, \quad I = 1, N. \qquad [5.84]$$

Variation	Equation	Comments
$\delta\dot{\varepsilon}_{ij}$ in A	$\sigma_{ij} = \Sigma_{ij}$	Proposed stresses are identified with constitutive stresses in the solid
$\delta\Sigma_{ij}$ in A	$\frac{1}{2}\left(\frac{\partial \dot{u}_i}{\partial X_j} + \frac{\partial \dot{u}_j}{\partial X_i}\right) = \dot{\varepsilon}_{ij}$	Compatibility enters the proposed strain rate and the proposed velocity in the solid
δr_i on S_u	$\dot{u}_i = \tilde{\dot{u}}_i$	Compatibility between the proposed velocity and the velocities imposed on the boundary of the solid
$\delta\dot{u}_i$ in A	$\frac{\partial \Sigma_{ji}}{\partial X_j} + F_i = 0$	Balance between the proposed stresses and the voluminal forces imposed in the solid
$\delta\dot{u}_i$ on S_t	$N_j\Sigma_{ji} = T_i$	Balance between the proposed stresses and the surface forces imposed on the boundary of the solid
$\delta\dot{u}_i$ on S_u	$N_j\Sigma_{ji} = r_i$	Balance between the proposed stresses and the support reactions proposed on the boundary of the solid

Table 5.12. *Euler equations for the materially nonlinear problems*

Moreover, the initial constitutive stresses (at the moment $t = 0$) are assumed null, and the initial values of the state variables q_{ij} are assumed constant in each Voronoi cell.

By time integration of the constitutive equations [5.82] and [5.83], it follows that at any moment, we obtain constant constitutive stresses in each Voronoi cell.

Consequently, each development of section 5.4 is easily transposed to the present case, and the discretized equations are similar to those obtained for the elastic linear case by means of replacing u_i with $\dot{u}_i$ and ε_{ij} with $\dot{\varepsilon}_{ij}$. They are summarized in Table 5.11 in matrix notation.

As a result, similar to the linear elastic case, we need not calculate the derivatives of the interpolation functions, and in the absence of body forces, no integration on the area of the cells is to be carried out. Only integrals along the contour of the Voronoi cells remain.

Equations	Comments	
$\{\sigma\}^I = \{\Sigma\}^I, I = 1, N$	Identification of the assumed stresses with the constitutive stresses in cell I	[5.85]
$\sum_{I=1}^{N} [A]^{IJ}\{\Sigma\}^I$ $- \sum_{K=1}^{M_u} B^{KJ}\{r\}^K$ $= \{\bar{F}\}^J + \{\tilde{T}\}^J, J = 1, N$	Equilibrium equation in cell J	[5.86]
$A_I\{\dot{\varepsilon}\}^I = \sum_{J=1}^{N} [A]^{IJ,T}\{\dot{u}\}^J,$ $I = 1, N$	Equation of compatibility in cell I	[5.87]
$\sum_{J=1}^{N} B^{KJ}\{\dot{u}\}^J = \{\dot{\bar{U}}\}^K,$ $K = 1, M_u$	Equation of compatibility on the side K subjected to imposed velocity	[5.88]

Table 5.13. *Discretized equations for materially nonlinear problems*

Considering that the material is inelastic, this result deserves to be highlighted.

5.12. Step-by-step solution of the discretized nonlinear equations

The discretized equations are solved by successive time steps. Although we are dealing with a known procedure, we offer the reader some details of the procedure to show that, despite the apparent number of unknowns due to the discretization, we ultimately get a system of equations with the same characteristics as in the classical approaches based on the discretization of the only displacements.

We assume that at time t_A, we know the values of all the useful variables:

$$^A\{\sigma\}^J, \ ^A\{\varepsilon\}^J, \ ^A\{\dot{u}\}^J, \ ^A\{u\}^J, \ ^A\{\tilde{F}\}^J, \ ^A\{\tilde{T}\}^J, \ ^A\{\tilde{u}\}^J, \ ^A\{r\}^J.$$

Thus these values verify all the equations in Table 5.11.

We want to obtain, at time $t_B = t_A + \Delta t$, the values $^B\{\sigma\}^J$, $^B\{\varepsilon\}^J$, $^B\{\dot{u}\}^J$, $^B\{r\}^J_i$.

Let

$$^B\{\sigma\}^J = {}^A\{\sigma\}^J + \{\Delta\sigma\}^J, \qquad\qquad [5.89]$$

$$^B\{u\}^J = {}^A\{u\}^J + \{\Delta u\}^J, \qquad\qquad [5.90]$$

During the interval Δt, it is admitted that nodal velocity is constant.

$$\{\dot{u}\}^J = \frac{^B\{u\}^J - {}^A\{u\}^J}{\Delta t} = \frac{\{\Delta u\}^J}{\Delta t}. \qquad\qquad [5.91]$$

It is therefore necessary to find the value of $^B\{u\}^J$, such that the equilibrium equations [5.86] are satisfied at the time t_B. We proceed by iterations.

Let $\{\Delta u\}^J_k$ be the value of $\{\Delta u\}^J$ at the iteration k and $\{\Delta u\}^J_{k+1}$ be the value of $\{\Delta u\}^J$ at the iteration $k + 1$.

We set the following:

$$\{\Delta u\}^J_{k+1} = \{\Delta u\}^J_k + \{du\}^J. \qquad\qquad [5.92]$$

From the equilibrium equation [5.86], we obtain:

$$\sum_{I=1}^{N} [A]^{IJ} \, {}^{B}\{\sigma\}_1^I - {}^{B}\{\tilde{F}\}^J + {}^{B}\{\tilde{T}\}^J - {}^{B}\{P\}_k^I = \{R\}_k^J \neq 0, \qquad [5.93]$$

$$\sum_{I=1}^{N} [A]^{IJ} \, {}^{B}\{\sigma\}_{k+1}^I - {}^{B}\{\tilde{F}\}^J + {}^{B}\{\tilde{T}\}^J - {}^{B}\{P\}_{k+1}^J = 0, \qquad [5.94]$$

where

$$\{P\}^J = \sum_{K=1}^{M_u} B^{KJ} \{r\}^K \qquad [5.95]$$

and ${}^{B}\{R\}_k^J$ are the out-of-balance forces at node J at the iteration k. From equations [5.93] and [5.94], we obtain:

$$\sum_{I=1}^{N} [A]^{IJ} \{d\sigma\}^I + \{R\}^J + {}^{B}\{P\}_k^J - {}^{B}\{P\}_{k+1}^J = 0, \qquad [5.96]$$

which provides the correction $\{d\sigma\}^I$, ensuring that the iteration $k+1$ is balanced. From the constitutive law, we can obtain:

$$\{d\sigma\}^I = [C_t]^I \{dd\dot{\varepsilon}\}^I \Delta t. \qquad [5.97]$$

$[C_t]^I$ is the consistent tangent compliance matrix, which must be evaluated according to the constitutive law of the material and the numerical scheme used for its time integration. Finally, from equations [5.87], [5.91], and [5.92], we obtain

$$\{d\dot{\varepsilon}\}^I A_I = \frac{1}{\Delta t} \sum_{I=1}^{N} [A]^{IJ,T} \{du\}^J. \qquad [5.98]$$

Equation [5.97] provides the link between the $\{du\}^J$ and the $\{d\dot{\varepsilon}\}^I$ which we need to determine. It is sufficient to insert equations [5.97] and [5.98] in equation [5.96] to lead to a system of the form:

$$[M]\{dq\} = \{dQ\}, \qquad [5.99]$$

where $[M]$ is given by equation [5.74] but with

$$[M]^{JL} = \sum_{I=1}^{N} [A]^{IJ} [C_t]^{I} [A]^{IL} \qquad [5.100]$$

$$\{dq\} = \begin{Bmatrix} \{du\}^1 \\ \{du\}^2 \\ . \\ . \\ . \\ \{du\}^N \end{Bmatrix} ; \qquad \{d\tilde{Q}\} = \{P\}_{k+1} - \{P\}_k + \{R\}_k. \qquad [5.101]$$

As in the linear elastic case, the $\{dq\}$ components corresponding to the imposed displacements are eliminated. We proceed in the same way for $[M]$ and $\{dQ\}$. Thus, we obtain a system of equations of the following form:

$$[M_p]\{dq_p\} = \{dQ_p\}, \qquad [5.102]$$

where all the components of $\{dQ_p\}$ are known. From this, we extract $\{dq_p\}$ and consequently, all the variables at the end of iteration $k + 1$. The iterations continue until convergence. For example, we can use the criterion of convergence as described below.

For an iteration k, we can calculate

$$\text{REF} = \text{REF}_1 + \text{REF}_2 \quad \text{with REF}_i = \frac{1}{N} \sum_{I=1}^{N} \left| \left({}^B\sigma_{ij}^I \right)_k A_j^{IJ} \right| \, i = 1, 2,$$
$$[5.103]$$

$$\text{RES} = \text{RES}_1 + \text{RES}_2 \quad \text{with RES}_i = \frac{1}{N} \sum_{I=1}^{N} \left| \left(R_i^I \right)_k \right| \, i = 1, 2, \qquad [5.104]$$

$$\text{RNORM} = \frac{\text{RES}}{\text{REF}}. \qquad [5.105]$$

If RNORM is lower than a sufficiently small prescribed value, we consider that convergence is attained. If the errors are weak, REF is approximately of the size of the interior nodal forces energetically equivalent to the stresses (of absolute value) in the solid, whereas RES measures the importance of the out-of-balance forces at iteration k.

In certain cases, for example, for a cyclic loading during which the stresses pass zero, REF can vanish. We can avoid this obstacle by defining a non-null value reference USER and by calculating REF by

$$\text{REF} = \max \left(\text{REF}_1 + \text{REF}_2, \text{USER} \right). \qquad [5.106]$$

5.13. Example of an elastoplastic material

To illustrate the preceding theory, here we consider the case of an elastoplastic material in plane stress state obeying the von Mises criterion with linear isotropic strain hardening.

The parameters describing the mechanical behavior of material are Young's modulus E, Poisson's ratio ν, plastic modulus h_p, and initial field limit R_e.

We use the radial return technique for stress calculations. The deviatoric stresses ${}^B\hat{\sigma}_{ij}$ at the end of the step are calculated by

$$ {}^B\{\hat{\sigma}\} = (1 - \beta)\{\hat{\sigma}^e\}, \qquad [5.107]$$

where $\{\hat{\sigma}^e\}$ is the elastic deviatoric stress obtained by assuming that the step is elastic

$$ \{\hat{\sigma}^e\} = {}^A\{\hat{\sigma}\} + 2G\{\dot{\hat{\varepsilon}}\}\Delta t, \qquad [5.108]$$

where $\{\dot{\hat{\varepsilon}}\}$ is the deviatoric strain rate.

In these formulae, we have

$$ G = \frac{E}{2(1+\nu)}; \qquad \beta = \frac{1 - \frac{R_e}{\sqrt{3}J_2^e}}{1 + \frac{h_p}{3G}}; \qquad J_2^e = \frac{\sqrt{\hat{\sigma}_{ij}^e \hat{\sigma}_{ij}^e}}{\sqrt{2}}. \qquad [5.109]$$

After some developments, we obtain the following equation:

$$ \{d\hat{\sigma}\} = 2G[\eta]\{d\hat{\sigma}^e\} = 2G[\eta]\{d\dot{\hat{\varepsilon}}\}\Delta t, \qquad [5.110]$$

where

$$ [\eta] = (1 - \beta) \begin{bmatrix} 1 & 0 & 0 \\ 0 & 1 & 0 \\ 0 & 0 & 1 \end{bmatrix} $$

$$ - \frac{R_e}{2\sqrt{3}(J_2^e)^3 \left(1 + \frac{h_p}{3G}\right)} \begin{bmatrix} \hat{\sigma}_{11}^e \hat{\sigma}_{11}^e & \hat{\sigma}_{11}^e \hat{\sigma}_{22}^e & 2\hat{\sigma}_{11}^e \hat{\sigma}_{12}^e \\ \hat{\sigma}_{22}^e \hat{\sigma}_{11}^e & \hat{\sigma}_{22}^e \hat{\sigma}_{22}^e & 2\hat{\sigma}_{22}^e \hat{\sigma}_{12}^e \\ \hat{\sigma}_{12}^e \hat{\sigma}_{11}^e & \hat{\sigma}_{12}^e \hat{\sigma}_{22}^e & \hat{\sigma}_{12}^e \hat{\sigma}_{12}^e \end{bmatrix}. \qquad [5.111]$$

The increment of the mean stress along the step is:

$$d\sigma_m = 3\chi d\dot{\varepsilon}_m \Delta t \qquad [5.112]$$

with

$$\chi = \frac{E}{3(1-2\nu)}; \qquad \sigma_m = \frac{1}{3}\sigma_{ii} = \frac{1}{3}(\sigma_{11} + \sigma_{22} + \sigma_{33}). \qquad [5.113]$$

By gathering the contributions of the deviatoric stress in equation [5.110] and of the mean stress in equation [5.112], we obtain:

$$\{d\sigma\} = \{d\hat{\sigma}\} + d\sigma_m\{\delta\}, \quad \text{with } \{\delta\} = \left\{\begin{array}{c} 1 \\ 1 \\ 0 \end{array}\right\}, \qquad [5.114]$$

which gives

$$\{d\sigma\} = [C_t]\{d\dot{\varepsilon}\}\Delta t \qquad [5.115]$$

with the consistent tangent compliance matrix

$$[C_t] = 2G[\eta] + \frac{1}{3}(3\chi[I] - 2G[\eta])\{\delta\}\{\delta\}^T, \qquad [5.116]$$

where $[I]$ represents the unit matrix.

5.14. Application: pure bending of an elastoplastic beam

Again we consider the example from section 5.9, but this time by applying some displacements imposed at the end of the right end as indicated in Figure 5.13. We define $u_{\min}$ and $u_{\max}$ (in mm) according to time by

$$|u_{\min}| = u_{\max} = t,$$

where t varies from 0 to 2 by increments $\Delta t = 0.02\,s$.

The curvature of the beam at time t is

$$\chi(t) = \frac{u_{\max}(t)}{hL}.$$

The constitutive law of material is that of section 5.13 with the following numerical values: $E = 200,000\,MPa$; $\nu = 0.3$; $h_p = 1,500\,MPa$; $R_e = 300\,Mpa$; $G = 76,923\,MPa$.

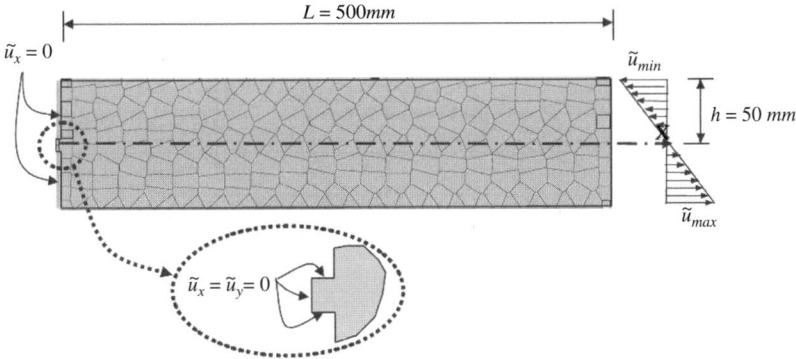

Figure 5.13. *Limit and loading conditions for the pure flexure of an elastoplastic beam*

The strain energy deduced by the mixed natural elements method is, at time t:

$$W_{I,\text{num}} = \sum_{I} A_I \int_{0}^{t} \langle \sigma^I \rangle \{\dot{\varepsilon}\}^I \, dt.$$

A direct solution for the plane stress state is calculated by directly integrating the differential equations of problem [ROS 07].

We note that W_0 as the corresponding strain energy. In this case, its value is:

$$W_0 = 4,265.11 \, J.$$

The values of $W_{I,\text{num}}$ at time $t = 2\,s$ for various numbers of nodes are given in Table 5.14.

Figure 5.14 compares the bending moment diagram obtained numerically with that obtained by the direct method [ROS 07].

Number of nodes N	$W_{I,\text{num}}\ (J)$
76	4,531.33
140	4,342.65
244	4,290.00

Table 5.14. *Convergence in energy, $W_0 = 4,265.11\ J$ at time $t = 2\ s$*

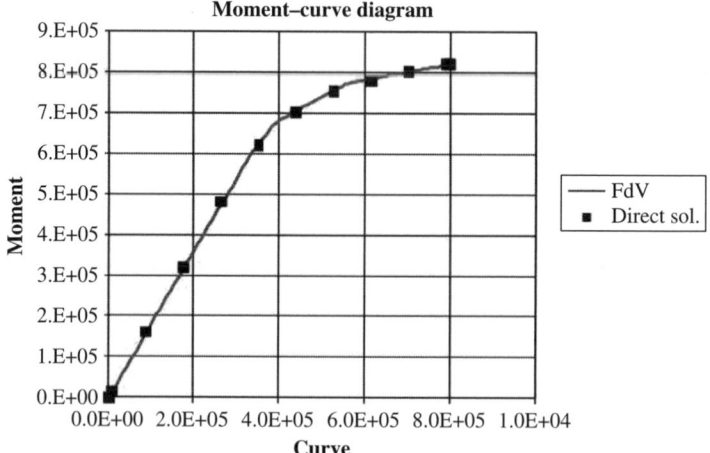

Figure 5.14. *Comparison of the bending-moment diagrams in monotonic loading*

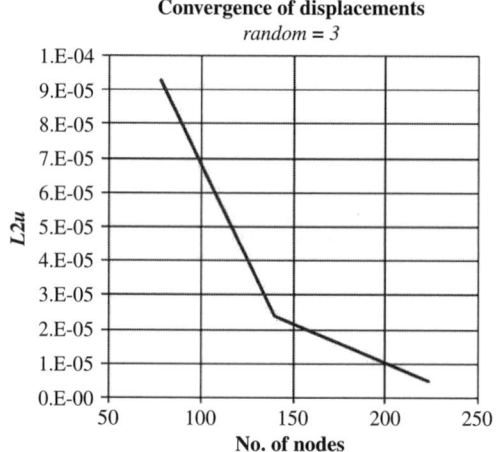

Figure 5.15. *Convergence in displacements*

Convergence in displacements was also evaluated using the norm

$$L2u = \frac{\sum_{i=1}^{N} \sqrt{\left(u_i^{th} - u_i^{\text{num}}\right) \times \left(u_i^{th} - u_i^{\text{num}}\right)}}{LN},$$

where u_i^{th} are the nodal displacements calculated analytically from the imposed curvature (which itself is deduced from $u_{\max}$) and u_i^{num} are those provided by the mixed natural elements method. Figure 5.15 illustrates this convergence when $t = 2\,s$.

Finally, an example of cyclic loading is given for the following history of loading:

$$\left|u_{\min}\right| = u_{\max} = \sin(t\pi), \quad 0 \le t < 1\,s,$$

$$\left|u_{\min}\right| = u_{\max} = 2\sin(t\pi), \quad 1 \le t < 2\,s,$$

$$\left|u_{\min}\right| = u_{\max} = 4\sin(t\pi), \quad 2 \le t < 4\,s,$$

where t varies by increments of $\Delta t = 0.01\,s$.

Figure 5.16 shows the bending moment diagram for this cyclic loading.

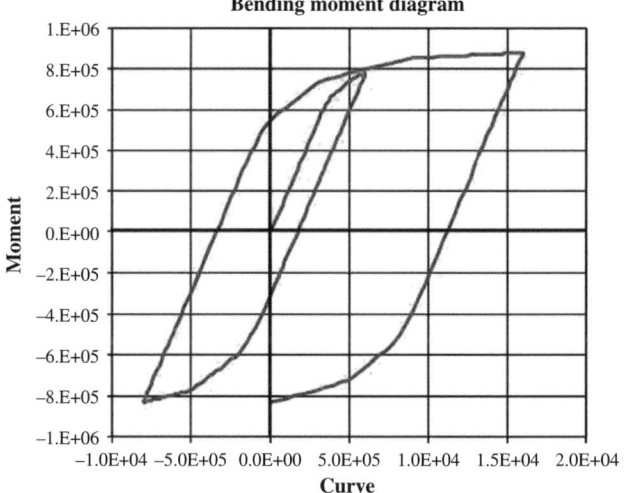

Figure 5.16. *Cyclic elastoplastic bending*

5.15. Conclusion

The mixed approach of the natural elements method based on the Fraeijs de Veubeke variational principle constitutes an alternative to the compatible approach based on the principle of virtual work.

In this approach, the stresses and the strains are assumed to be constant in each Voronoi cell. The support reactions are assumed to be constant on each side of the Voronoi cells, and the displacements are interpolated using Laplace or Sibson functions as in the compatible approach.

The mixed approach (in the linear elastic domain or in the small inelastic strains domain) has the following characteristics:

– There is no need to calculate the derivatives of the interpolation functions.

– In the absence of body forces, there is no numerical integration on the area of the studied domain: only integrals along the sides of the Voronoi cells are necessary, for which a Gauss scheme with two integration points for each side is sufficient.

– It is possible to impose any displacements on a part of the studied contour domain. On average, they are respected on each side of the Voronoi cells.

– In spite of the flexibility of the discretization, which introduces a large number of unknowns from the beginning, ultimately we come back to a classical symmetric system of equations, having only nodal displacements as unknowns.

– The patch test is respected up to machine precision.

– There is no numerical locking when the material is quasi-incompressible.

Chapter 6

Flow Models

The possibility of letting the nodes flow with material velocity without having to worry about the quality of the underlying mesh opens up a new possibility in the simulation of flow models. It is then possible to follow an updated Lagrangian approach. In this chapter, various applications that can prove the potentiality of the above-mentioned procedure are discussed.

6.1. Natural element method in fluid mechanics: updated Lagrangian approach

In general, meshless methods are well adapted to problems that present large transformations where the finite elements techniques have the drawback of needing frequent remeshing. This section considers some examples of Newtonian or non-Newtonian fluid-flow simulations. The first application (to our knowledge) of the natural element method (NEM) in fluid mechanics is found in [BRA 95a]. Then, this method was applied to the Newtonian and non-Newtonian fluid-flow simulations (in particular microstructural fluids) [MAR 02], and was compared with the techniques with a fixed mesh in [MAR 03].

6.1.1. *Mechanical model of a Newtonian fluid flow*

The problem is defined by equations [6.1]–[6.3]:

– Equilibrium (in absence of terms of mass and inertia):

$$\nabla \cdot \boldsymbol{\sigma} = \mathbf{0}. \qquad [6.1]$$

– Incompressibility:

$$\nabla \cdot \boldsymbol{v} = 0. \qquad [6.2]$$

– Constitutive law of a Newtonian fluid:

$$\sigma = -p\boldsymbol{I} + 2\mu\boldsymbol{d}, \tag{6.3}$$

where μ represents the viscosity of the fluid, $\boldsymbol{I}$ the unit tensor, and $\boldsymbol{d}$ the strain rate tensor.

The associated variational formulation can be written as: to find the kinematically admissible velocity field $v \in \mathcal{U}$ such as:

$$\int_{\Omega(t)} \sigma : \boldsymbol{d}^* \, \mathrm{d}\Omega = \int_{\Gamma_t} \bar{\boldsymbol{t}} \cdot \boldsymbol{v}^* \mathrm{d}\Gamma \quad \forall \boldsymbol{v}^* \in \mathcal{V} \tag{6.4}$$

$$\int_{\Omega(t)} (-\nabla \boldsymbol{v} + \varepsilon p) p^* \mathrm{d}\Omega = 0 \quad \forall p^* \in L_2(\Omega(t)), \tag{6.5}$$

where $\mathcal{U} = \{v | v \in \left(H^1(\Omega(t))\right)^2, \ v|_{\Gamma_v} = \bar{v}\}$, $\mathcal{V} = \{v^* | v^* \in \left(H^1(\Omega(t))\right)^2, \ v|_{\Gamma_v} = 0\}$, and Γ_v is the part of the domain boundary where the velocities are prescribed. H^1 and L_2, respectively, are Sobolev and Lebesgue function spaces, and ϵ is a penalty parameter. To estimate velocities and pressures, we consider a mixed interpolation of Sibson–Thiessen $C^0 - C^{-1}$.

6.2. Free and moving surfaces

A detailed study in [MAR 03] presents domain shape extraction as it evolves through time. For complex free surface flows, the explicit follow-up of the domain boundary is costly in terms of calculation. In [MAR 03], we proposed the use of alpha shapes to extract the domain shape from the node cloud, and therefore, also boundary, using only the knowledge of the node clouds at each moment. At every moment, a suitable alpha shape is used to solve equations [6.4] and [6.5].

The correct alpha value must be chosen that depends on the average distance between nodes h, on the desired level of detail in the geometrical representation (section 2.3.1), as well as on the time steps considered in the temporal evolution.

To illustrate these reflections, we study the case of a fluid that runs between two parallel plates where a circular obstacle is placed (Figure 6.1). A certain number of nodes with an imposed velocity $v_x(x = 0, y) = \frac{4V}{h^2}(h - y)y$ ($V = 0.1$) are injected into the domain at each time step.

As already specified (section 2.3.1) to fully represent the geometry, the alpha parameter must be chosen so as to correctly represent the smallest radius of curvature.

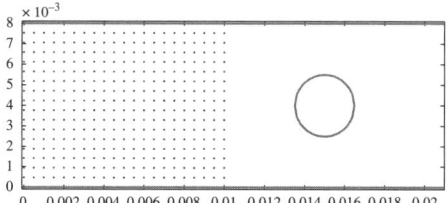

Figure 6.1. *Flow problem between two parallel plates with an obstacle*

In this case, alpha is thus restricted by the radius of obstacle $R = 0.0015$. If alpha is larger than this value, the domain cannot be represented correctly (Figure 6.2(a)).

From Figure 6.2, we can conclude that a wrong choice for the alpha value produces significant errors in the mass conservation during simulation. Figure 6.3 compares the volume of fluid injected into the domain with that represented by the alpha shape for different alpha values. If the alpha value is too big, it results in a value that is consequently too big for the volume (considering the obstacle as already encountered in the fluid domain in Figure 6.2(a)). However, if the alpha value is too small, then the extreme distorted triangles (even if its presence does not affect the accuracy of the calculations) are not considered by the alpha shape because the radius of a circle circumscribed to these triangles becomes too large compared with the alpha value. For alpha values in the interval between the h and the radius of the obstacle, the results appear totally satisfactory with regard to the representation of the geometry:

$$h < \alpha < R. \tag{6.6}$$

Another source of errors is associated with the considered time steps. Thus, if Δt is too large, $|v|\Delta t > h$, the particles will move a distance higher than the average spacing of the nodes h during each time step. If two flow fronts meet each other (downstream from the obstacle for example), each front would penetrate into the opposite fluid domain. Consequentially, this results in the loss of a part of the fluid represented by the alpha shape. Moreover, if the $alpha$ value is too large, the two fronts will be welded even before their collision.

We can therefore conclude that the conservation of the mass imposes $\mathcal{O}(\alpha) \approx \mathcal{O}(h)$, leading to the resulting time steps:

$$\Delta t = \min_{x} \left\{ \frac{\alpha(x, t)}{|v(x, t)|} \right\}. \tag{6.7}$$

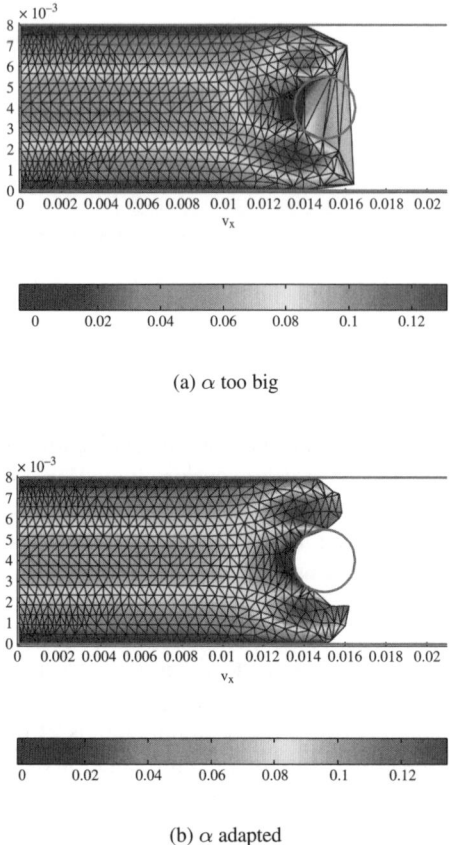

(a) α too big

(b) α adapted

Figure 6.2. *Effect of the alpha value on the representation of geometry*

6.2.1. *Use of the characteristics method*

The simulations presented here are carried out within the framework of an updated Lagrangian formulation at each time step. The nodes are transported by material velocity. By taking the evolution of inner variables into account, which is usually controlled by transport equations, this can be done with a great accuracy by using the characteristics method. Moreover, the same strategy can be applied for dealing with the convection terms of the models within a updated Lagrangian formulation framework with the decomposition of the operator in an elliptic part and a hyperbolic part.

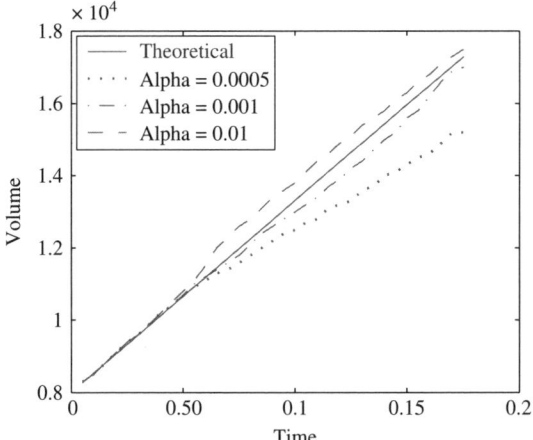

Figure 6.3. *Injected volume versus represented volume for various alpha
parameter values*

We consider now the case of a fluid under the effects of gravity:

$$\boldsymbol{\nabla} \cdot \boldsymbol{\sigma} + \rho \boldsymbol{b} = \rho \frac{d\boldsymbol{v}}{dt} = \rho \left(\frac{\partial \boldsymbol{v}}{\partial t} + \boldsymbol{v}\boldsymbol{\nabla}\boldsymbol{v} \right) \qquad [6.8]$$

$$\boldsymbol{\nabla}\boldsymbol{v} = 0 \qquad [6.9]$$

$$\boldsymbol{\sigma} = -p\boldsymbol{I} + \boldsymbol{\tau} = -p\boldsymbol{I} + 2\mu\boldsymbol{d}, \qquad [6.10]$$

where $\boldsymbol{b} = (0, 0, -g)$ defines the acceleration of gravity, μ the viscosity of the fluid,
$\boldsymbol{d}$ the stress rate tensor, and $\boldsymbol{\tau}$ the extra-strain tensor. If we insert the constitutive law
(equation [6.10]) into the conservation equation of linear momentum [6.8] we obtain:

$$\boldsymbol{\nabla} \cdot \boldsymbol{\tau} - \boldsymbol{\nabla} p + \rho \boldsymbol{b} = \rho \frac{d\boldsymbol{v}}{dt} \qquad [6.11]$$

The variational formulation of the problem is then:

$$\int_{\Omega} 2\mu\boldsymbol{d} : \boldsymbol{d}^* d\Omega - \int_{\Omega} p\boldsymbol{I} : \boldsymbol{d}^* d\Omega = -\int_{\Omega} \rho \boldsymbol{b}\boldsymbol{v}^* d\Omega + \int_{\Omega} \rho \frac{d\boldsymbol{v}}{dt} \boldsymbol{v}^* d\Omega \qquad [6.12]$$

$$\int_{\Omega} \boldsymbol{\nabla} \cdot \boldsymbol{v} p^* d\Omega = 0. \qquad [6.13]$$

The last term of the second member of equation [6.12] can be integrated along the
trajectories using the method of characteristics. Thus, if we propose that the solution
is known at the time step $t_{n-1} = (n-1)\Delta t$, we search for the solution at step t_n

$$\int_{\Omega} \rho \frac{d\boldsymbol{v}}{dt} \boldsymbol{v}^* = \int_{\Omega} \rho \frac{\boldsymbol{v}^n(\boldsymbol{x}) - \boldsymbol{v}^{n-1}(\boldsymbol{X})}{\Delta t} \boldsymbol{v}^* d\Omega, \qquad [6.14]$$

where X is the position of the particle at time t_{n-1} of the point that occupies the position x at time t_n previous time step t_{n-1}:

$$x = X + v^{n-1}(X)\Delta t, z^{\iota} \qquad [6.15]$$

finally giving

$$\int_{\Omega} 2\mu d : d^*\mathrm{d}\Omega - \int_{\Omega} pI : d^*\mathrm{d}\Omega - \int_{\Omega} \rho\frac{vv^*}{\Delta t} = -\int_{\Omega} \rho bv^*\mathrm{d}\Omega - \int_{\Omega} \rho\frac{v^{n-1}v^*}{\Delta t}\mathrm{d}\Omega \qquad [6.16]$$

$$\int_{\Omega} \nabla \cdot vp^*\mathrm{d}\Omega = 0, \qquad [6.17]$$

where equation [6.16] alone deserves more comments. If we use a stabilized nodal integration, the only important aspect is the need to store the nodal velocity at time t_{n-1}.

6.3. Short-fiber suspensions flow

Recently, we tried to combine the NEM (within the framework of updated Lagrangian formulations) with a model reduction technique for the simulation of complex fluids with microstructure (short-fiber suspensions).

The suspension model couples the flow kinematics with the evolution of the microstructure. Indeed, the existence of this microstructure affects the mode of flow, and it is precisely the kinematics of the flow, which control the evolution of the microstructure (fibers). In general, the two models are coupled with a fixed-point strategy (in the stationary case) or with an explicit strategy in the case of evolutionary models. Fully coupled models are rather rare, due to the various characters in the momentum equations and fiber orientations, as well as due to the difficulty of treatment of the nonlinearity.

Thus, we are confronted with the solution of a nonlinear convection equation, which controls the fibers. Being given the hyperbolic character of the aforementioned method, various stabilizations have been used within the framework of eulerian simulations (fixed mesh): SUPG, Taylor Galerkin discontinuous method, characteristics method, etc.

In [CHI 05], a new approach was proposed by combining a solution of the flow model starting from the NEM and a model reduction method on a microscopic scale to reduce the degrees of freedom, and consequently the calculation time, during the calculation stage of the evolution of orientation.

6.3.1. *Flow kinematics*

The flow model of a short-fiber suspension is given by the equilibrium equation [6.1] and the incompressibility equation [6.2]. The resultant constitutive law is written in the form:

$$\sigma = -p\boldsymbol{I} + 2\mu\{\boldsymbol{d} + N_p(\boldsymbol{a}_4 : \boldsymbol{d})\}, \qquad [6.18]$$

where N_p is a parameter whose value depends on the fiber concentration as well as on the aspect ratio. $\boldsymbol{a}$ is the fourth-order orientation tensor, which is defined as

$$\boldsymbol{a}_4 = \int \boldsymbol{\rho} \otimes \boldsymbol{\rho} \otimes \boldsymbol{\rho} \otimes \boldsymbol{\rho}\psi(\boldsymbol{\rho})\mathrm{d}\boldsymbol{\rho} \qquad [6.19]$$

where $\boldsymbol{\rho}$ is a unit vector aligned with the fiber, and $\psi(\boldsymbol{\rho})$ the distribution function of the orientations, whose evolution is controlled by the Fokker–Planck equation.

If we consider that $\psi(\boldsymbol{\rho}) = \delta(\psi(\boldsymbol{\rho}) - \psi(\overline{\boldsymbol{\rho}}))$, with δ, the Dirac delta distribution, all the probability of orientation will be concentrated in a direction defined by $\overline{\boldsymbol{\rho}}$. It is common to define the second order orientation tensor **a** as:

$$\boldsymbol{a} = \int \boldsymbol{\rho} \otimes \boldsymbol{\rho}\psi(\boldsymbol{\rho})\mathrm{d}\boldsymbol{\rho}. \qquad [6.20]$$

Sometimes, we express the tensor of fourth order according to that of the second order. In general, these expressions are not exact, and they are known as closure relations.

If we consider ellipsoidal shape fibers, the evolution of their orientations is given by the Jeffery expression (valid in the case of diluted suspensions):

$$\frac{\mathrm{d}\boldsymbol{\rho}}{\mathrm{d}t} = \boldsymbol{\Omega}\boldsymbol{\rho} + k\Big(\boldsymbol{D}\boldsymbol{\rho} - (\boldsymbol{d} : (\boldsymbol{\rho} \otimes \boldsymbol{\rho}))\boldsymbol{\rho}\Big), \qquad [6.21]$$

where $\boldsymbol{\Omega}$ is the vorticity tensor, and k a parameter which depends on aspect ratio of the fibers. The evolution of the orientation distribution function is given by the equation of Fokker–Planck:

$$\frac{\mathrm{d}\psi}{\mathrm{d}t} + \frac{\partial}{\partial\boldsymbol{\rho}}\Big[\psi\frac{\mathrm{d}\boldsymbol{\rho}}{\mathrm{d}t}\Big] = 0. \qquad [6.22]$$

By combining the preceding equations, we also obtain the evolution of the second order orientation tensor:

$$\frac{\mathrm{d}\boldsymbol{a}}{\mathrm{d}t} = \boldsymbol{\Omega}\boldsymbol{a} - \boldsymbol{a}\boldsymbol{\Omega} + k\Big(\boldsymbol{d}\boldsymbol{a} + \boldsymbol{a}\boldsymbol{d} - 2(\boldsymbol{a}_4 : \boldsymbol{d})\Big). \qquad [6.23]$$

6.3.2. *Coupling of a particle method with α-NEM*

We consider, in a 2D case, N particles (pseudo-fibers) with an initial isotropic distribution. These orientations can be defined by

$$\theta_i = i \times \frac{2\pi}{N}, \quad i \in [0, \dots, N-1]. \qquad [6.24]$$

In 2D, the orientation is expressed by

$$\rho = \begin{pmatrix} \cos\theta \\ \sin\theta \end{pmatrix}, \qquad [6.25]$$

which is inserted into the Jeffery equation, and after some simple manipulation, leads to

$$\dot{\theta} = \begin{pmatrix} -\sin\theta & \cos\theta \end{pmatrix} \begin{pmatrix} 0 & \frac{1}{2}\left(\frac{\partial u}{\partial y} + \frac{\partial v}{\partial x}\right) \\ -\frac{1}{2}\left(\frac{\partial u}{\partial y} + \frac{\partial v}{\partial x}\right) & 0 \end{pmatrix} \begin{pmatrix} \cos\theta \\ \sin\theta \end{pmatrix}$$

$$+ k \begin{pmatrix} -\sin\theta & \cos\theta \end{pmatrix} \begin{pmatrix} \frac{\partial u}{\partial x} & \frac{1}{2}\left(\frac{\partial u}{\partial y} + \frac{\partial v}{\partial x}\right) \\ \frac{1}{2}\left(\frac{\partial u}{\partial y} + \frac{\partial v}{\partial x}\right) & \frac{\partial v}{\partial y} \end{pmatrix} \begin{pmatrix} \cos\theta \\ \sin\theta \end{pmatrix}. \qquad [6.26]$$

In this manner, the velocity of fiber rotation depends on θ fiber orientation, the gradient velocities, and the ratio of fibers. Figure 6.4 shows the field of orientation – the velocity and the associated alpha shape – at one instant of an extrusion process.

6.4. Breaking dam problem

Here we deal with a common example (see [BON 00, RAM 87, LEW 97, RAB 03]). In this problem, one of the walls of a rectangular tank filled with a fluid is removed instantaneously, thus allowing the gravity to trigger the flow of the liquid (Figure 6.5).

The dimensions of the tank are 0.05715 m in height and 0.05715 m in length. We considered in our simulations a slip condition between the fluid and various solid surfaces, as well as density of the fluid of 10^3 kg/m^3 and dynamic viscosity of 0.1 kg · m^{-1}s^{-1}.

Figure 6.6 compares the numerical prediction of the position of the flow front over time with the experimental measurements published in [MAR 52]. In this figure, all dimensions were adimensionalized compared with the length of the tank a and with time $t\sqrt{g/a}$. In the Y-axis we plot z/a, where z is the position of the front over time.

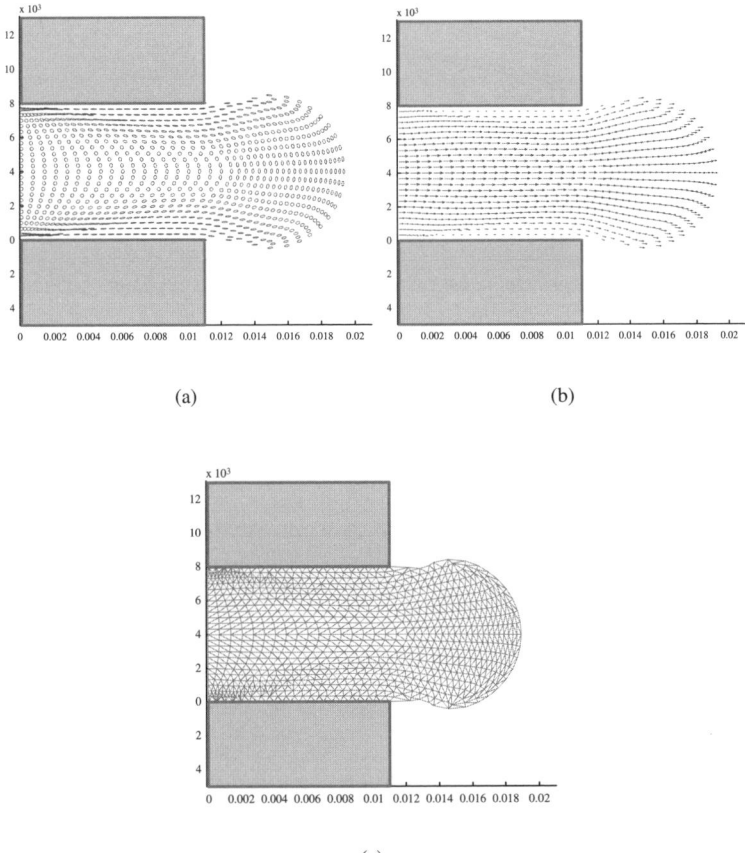

Figure 6.4. *Simulation of extrusion: orientation (a), velocities (b) and alpha shape (c)*

The error relating to the mass conservation is represented in Figures 6.9 and 6.10. This error is lower than 3%, which proves the high accuracy of the alpha shapes to extract the geometry from the domain fluid with the only information relative to the nodal distribution. The three velocity fields are represented in Figure 6.11.

6.5. Multi-scale approaches

These techniques, which couple the molecular scales of the kinetic theory with the macro scale of the continuous medium, have received increasing interest in the last few years [LAS 93, GIG 02], in spite of a very significant numerical cost. The

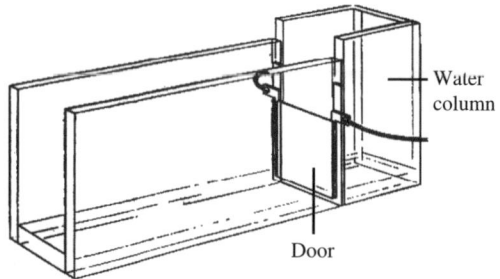

Figure 6.5. *Experimental construction associated with the problem of a dam rupture (according to [MAR 52])*

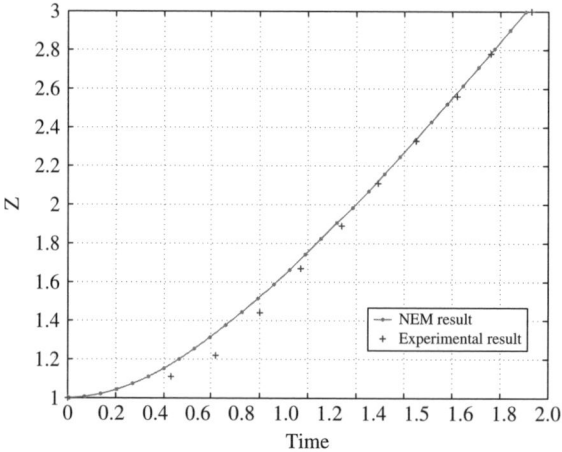

Figure 6.6. *Comparison between the numerical prediction of the position of the flow front and experimental measurements*

advantage of this type of approach comes from the equivalence between the Fokker–Planck equation, which describes the evolution of the probability of the distribution of a micro-scale configuration and a stochastic differential equation of Itô. Thus, instead of solving the Fokker–Planck deterministic equation, an equivalent stochastic differential equation is solved by integrating the sufficient number of trajectories of the stochastic process. The viscoelastic stresses are obtained by carrying out averages of the stochastic process.

In the original version of the method known as CONNFFESSIT, simulation starts by solving the conservation equations, usually by Eulerian or arbitrary lagrangian eulerian (ALE) methods, and by employing finite elements. With the field velocity updated, a group of molecules is transported in the flow. Its trajectory is described in

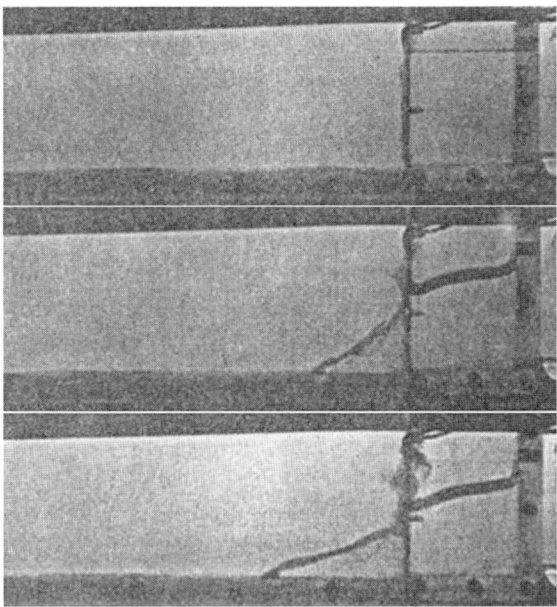

Figure 6.7. *Visualization of the dam rupture (according to [MAR 52])*

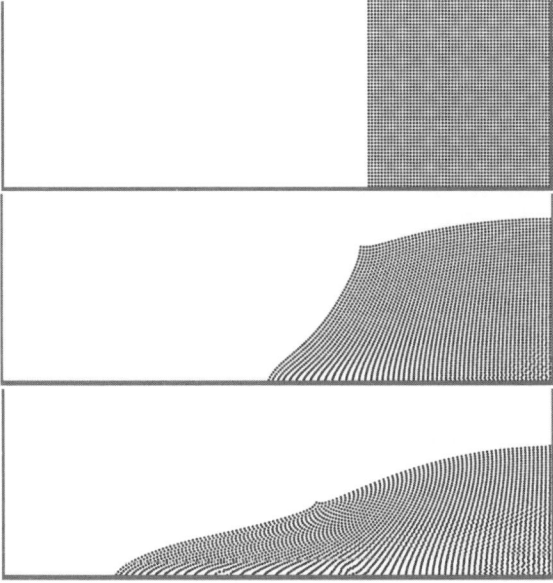

Figure 6.8. *Numerical simulation of the dam rupture*

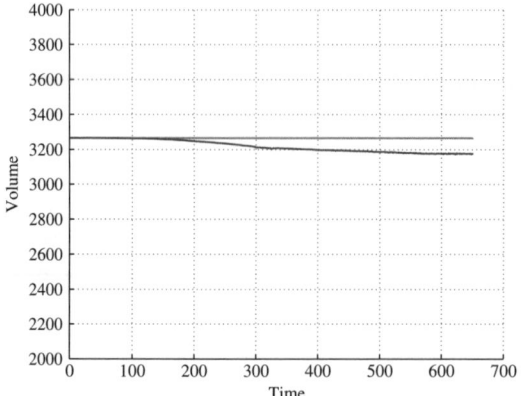

Figure 6.9. *Error in the mass conservation–evolution of the mass*

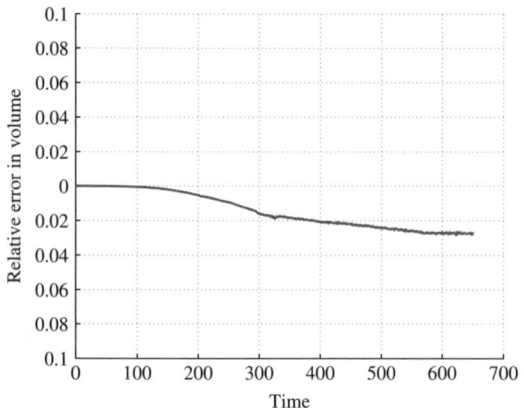

Figure 6.10. *Error in the mass conservation–relative Error*

a Lagrangian way. The equation of Itô is then integrated along each trajectory, and finally the viscoelastic stresses are obtained by an average of the configuration state of the particles in each finite element.

Despite its success in many applications, this approach presents some disadvantages (see [KEU 04] for example). First, the molecules must be localized within the finite element mesh. Second, this search proves to be very costly in the case of complex geometries. Finally, the number of particles placed in each element

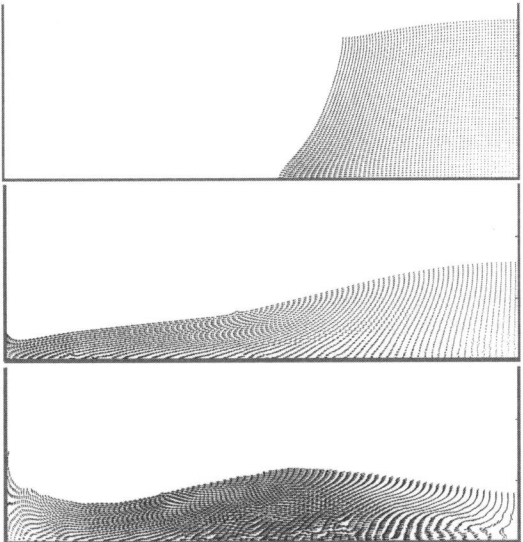

Figure 6.11. *Velocity fields at three moments*

of the finite elements mesh must be large enough to limit the stochastic noise during the estimate of the viscoelastic stress.

These problems can be avoided within the framework of a Lagrangian approach. The molecules are then attached to each node, and they follow the material velocity. In this way, the search for the element in which the particle is located becomes ineffectual. The problem traditionally associated with the Lagrangian approaches is the distortion of the mesh and the numerical diffusion of the results remeshing associated with it. The family of meshless methods [BEL 94] has become famous precisely due to the absence of remeshing. Thus, meshing is not needed, and the connectivity between the nodes is managed by the method without the intervention of a particular algorithm.

The meshless methods have opened up the possibility of employing Lagrangian approaches in several branches of computational mechanics [MAR 03, MAR 04, ALF 06b, GON 07a]. In this section, we analyze the contribution of these methods in solving the problems associated with the micro-macro models for viscoelastic flows.

In the examples that follow, we employed the NEM, which offers some advantages for these kinds of applications. The imposition of boundary conditions, for example, is direct by employing α-NEM [CUE 00] or constrained natural element method

[YVO 04b].   Monitoring the free surface of the flow is also direct.   The use of techniques of the Volume of Fluid type (VoF) [BON 06] can also be avoided.

6.5.1. *Mechanical model*

According to Keunings [KEU 04], the kinetic theory has three main ingredients. The first is the configuration space, represented here by X. It represents the variables necessary to represent the molecular configuration. The second ingredient characterizes the state of a microstructure through the function of probability distribution $\psi(X, x, t)$. In balance, the function of probability distribution is noted by ψ_{eq}. Finally, the third ingredient is the relation that makes it possible, starting from the function of distribution ψ, to obtain the extra stresses τ_p that are viscoelastic stresses which appear in non-equilibrium situations.

The field velocities and stresses are governed by the following equations:

$$\nabla \sigma = 0 \qquad [6.27]$$

$$\nabla v = 0 \qquad [6.28]$$

$$\sigma = -pI + \tau + 2\eta\, \dot{\gamma}, \qquad [6.29]$$

where $\dot{\gamma}$ represents the symmetric part of the gradient of velocities tensor $\dot{\gamma} = \frac{1}{2}(\nabla v + \nabla v^T)$, η the viscosity of solvent (if it exists), and τ the contribution of the polymer to the stresses at a microscopic level. The resulting model couples two scales in a nonlinear way.

The simplest algorithm to solve this problem consists of using an explicit strategy, which makes it possible to linearize the microscopic model by decoupling the two scales from the description. Thus, the variational form related to equations [6.27] and [6.29] is written as:

$$\int_\Omega 2\eta\, \dot{\gamma}^n : \dot{\gamma}^* d\Omega - \int_\Omega p^n I : \dot{\gamma}^* d\Omega + \int_\Omega \tau^{n-1} : \dot{\gamma}^* d\Omega = 0. \qquad [6.30]$$

To avoid the loss of ellipticity associated with the absence of solvent contribution, that is, $\eta = 0$ in equation [6.30], as is often the case for melted polymer models, the simplest method consists of adding a viscous term on both the sides of equation [6.30] (at different time steps to avoid their direct cancellation):

$$\begin{aligned} \int_\Omega 2\eta\, \dot{\gamma}^n : \dot{\gamma}^* d\Omega - \int_\Omega p^n I : \dot{\gamma}^* d\Omega + \int_\Omega 2\mu\, \dot{\gamma}^n : \dot{\gamma}^* d\Omega \\ = -\int_\Omega \tau^{n-1} : \dot{\gamma}^* d\Omega + \int_\Omega 2\mu\, \dot{\gamma}^{n-1} : \dot{\gamma}^* d\Omega \end{aligned} \qquad [6.31]$$

which, with the equation linked to incompressibility:

$$\int_{\Omega(t)} (-\mathrm{Div}\boldsymbol{v}^n + \varepsilon p^n)p^* d\Omega = 0, \tag{6.32}$$

completes the macroscopic description of the problem. It should be noted that the incompressibility is imposed in our approach by the addition of a penalization parameter ε; however, more sophisticated approaches exist, such as DEVSS [BAA 98].

To finish the formulation, the kinetic theory also gives an equation that controls the evolution of the probability density function ψ. This equation is known as the Fokker–Planck equation:

$$\frac{\mathrm{D}\psi}{\mathrm{D}t} = -\frac{\partial}{\partial \boldsymbol{X}} \cdot \{\boldsymbol{A}\psi\} + \frac{1}{2}\frac{\partial}{\partial \boldsymbol{X}}\frac{\partial}{\partial \boldsymbol{X}} : \{\boldsymbol{D}\psi\}, \tag{6.33}$$

where $\boldsymbol{A}$ represents a vector that describes the drift exerted by the flow on the function ψ, and $\boldsymbol{D}$ is a symmetrical matrix, positive definite, which describes the Brownian effects in the model. $\mathrm{D}/\mathrm{D}t$ represents the material derivative. The expression which finally binds the state of the configuration with the stresses within the fluid is expressed by

$$\boldsymbol{\tau} = \int \boldsymbol{g}(\boldsymbol{X})\psi d\boldsymbol{X} = \langle \boldsymbol{g}(\boldsymbol{X})\rangle, \tag{6.34}$$

where the hooks $\langle\cdot\rangle$ indicate an average on the whole of the conformation space in a physical point. $\boldsymbol{g}$ is a function defined on the current configuration, which depends on the considered model. In spite of the existence of new deterministic techniques to solve equation [6.33] [AMM 06b, AMM 06a, AMM 07], the stochastic simulation strategies are widely used.

The stochastic approach of the problem benefits from the equivalence between the Fokker–Planck [6.33] equation and its associated stochastic differential equation [ÖTT 96]:

$$\mathrm{d}\boldsymbol{X} = \boldsymbol{A}\mathrm{d}t + \boldsymbol{B} \cdot \mathrm{d}\boldsymbol{W}, \tag{6.35}$$

where

$$\boldsymbol{B} \cdot \boldsymbol{B}^T = \boldsymbol{D}, \tag{6.36}$$

and $\boldsymbol{W}$ represents a Wiener process. Equation [6.35] is applied to each individual trajectory of the molecules. Thus, in this approach, these molecules are attached to the nodes, which makes it possible to carry out the integration of equation [6.35] by the characteristics method. Integration is carried out by the Euler–Maruyama method:

$$\boldsymbol{X}_{n+1}^j = \boldsymbol{X}_n^j + \boldsymbol{A}(\boldsymbol{X}_n^j, t_n)\Delta t + \boldsymbol{B}(\boldsymbol{X}_n^j, t_n) \cdot \Delta \boldsymbol{W}_n^j, \tag{6.37}$$

where n represents the current time step and j represents the considered molecule.

We now present two different models based on the Fokker–Planck approach, the first of which is the finite extension nonlinear elastic (FENE) model for the dissolution of polymers. The second is the model by Doi and Edwards for entangled polymers, which is based on the concept of reptation.

6.5.1.1. *FENE model*

The FENE model considers the polymer chains as a group of elastic dumbbells which do not interact, and immersed in a Newtonian medium. Each dumbbell is made up of two particles connected by a spring. Thus, the conformation space is expressed by $X = Q$, where Q is the vector that binds the two particles of the dumbbell. We postulate that the spring has nonlinear behavior. This comes from the fact, among other, that the spring has a limited extension Q_0. The equation representing the behavior for this spring can be written, according to [OWE 02]:

$$F^c = \frac{H}{1 - Q^2/Q_0^2} Q, \qquad [6.38]$$

with H as the constant of the spring. Each particle is subject to the action of the hydrodynamic forces given by:

$$F^h = -\zeta \left[\langle \dot{r} \rangle - (\kappa \cdot r + v_0) \right], \qquad [6.39]$$

where r represents the position of each particle, κ the velocity gradient tensor, and $\kappa \cdot r + v_0$ the field velocity of the fluid on the scale of the dumbbell (here, it is estimated that the strain field is homogeneous on the scale of the molecule). Finally, ζ represents the viscous friction coefficient between the particles and the solvent. However, the molecules are subjected to Brownian forces due to the frequently numerous collisions with the solvent molecules:

$$F^b = -kT \left(\frac{\partial \ln \Psi}{\partial r} \right), \qquad [6.40]$$

The non-dimensional stress tensor for the FENE model is given by

$$\tau = -nk_B T \left\langle \frac{\hat{Q}\hat{Q}}{1 - \frac{\hat{Q}^2}{b}} \right\rangle + I, \qquad [6.41]$$

where $\hat{Q} = Q/(k_B T/H)^{1/2}$ and k_B represents the Boltzmann constant. $\hat{t} = t/\lambda_H$ with $\lambda_H = \zeta/4H$ is a time which is characteristic of a polymer. b represents the adimensional parameter of limit.

Non-dimensional equation for this model [HER 97] is finally written as:

$$\frac{\partial \hat{\psi}}{\partial \hat{t}} = -\frac{\partial}{\partial \hat{Q}} \cdot \left(\kappa \cdot \hat{Q} \, \hat{\psi} - \frac{1}{2} \frac{1}{1 - \hat{Q}^2/b} \hat{Q}\hat{\psi} \right) + \frac{1}{2} \frac{\partial}{\partial \hat{Q}} \cdot \frac{\partial}{\partial \hat{Q}} \hat{\psi}. \qquad [6.42]$$

In this last equation, $\hat{\kappa}$ represents the adimensional form of the velocity gradient. The function of distribution is finally

$$\hat{\psi}(\hat{Q}, \hat{t}) = \frac{\psi(Q, t)}{\left(\frac{k_B T}{H}\right)^{\frac{3}{2}}}.$$ [6.43]

The equivalent differential equation of Itô for this model, equation [6.35], is given by

$$d\hat{Q}_{\hat{t}} = \left(\hat{\kappa} \cdot \hat{Q} - \frac{1}{2} \frac{1}{1 - \hat{Q}_{\hat{t}}/b} \hat{Q}_{\hat{t}}\right) d\hat{t} + d\boldsymbol{W}_{\hat{t}}.$$ [6.44]

In this case, $\boldsymbol{W}_{\hat{t}}$ is a three-dimensional Wiener process. To numerically integrate this equation, a semi-implicit predictor-corrector method of second order proposed by [ÖTT 96] was employed. This method was also employed in [HER 97].

6.5.1.2. Doi–Edwards model

The reptation model considered here is the well-known model by Doi–Edwards or Curtiss–Bird [ÖTT 96]. The configuration space $\boldsymbol{X}$ is defined by a vector of orientation of the segment of tube $\boldsymbol{u}$ and by the standardized position of the chain in the tube, $s \in [0, 1]$. Thus, the probability distribution function φ is related to $\boldsymbol{u}$, s, $\boldsymbol{x}$, and t. The expression $\psi(\boldsymbol{u}, s, \boldsymbol{x}, t) d\boldsymbol{u} ds$ represents the probability of finding a tube segment at time t and in position $\boldsymbol{x}$, oriented between $\boldsymbol{u}$ and $\boldsymbol{u} + d\boldsymbol{u}$ which contains a polymer chain segment $[s, s + ds]$. The associated Fokker–Planck equation is then written as:

$$\frac{D\psi}{Dt} = -\frac{\partial}{\partial \boldsymbol{u}} \cdot [(\boldsymbol{I} - \boldsymbol{uu}) \cdot \boldsymbol{\kappa} \cdot \boldsymbol{u}\psi] + \frac{1}{\pi^2 \tau_d} \frac{\partial^2 \psi}{\partial s^2},$$ [6.45]

with τ_d the time reptation [KEU 04]. The constitutive law used is

$$\boldsymbol{\tau} = G\langle \boldsymbol{uu} \rangle.$$ [6.46]

The Fokker–Planck equation now is equivalent to two stochastic differential equations of Itô. The first equation is deterministic

$$d\boldsymbol{u} = (\boldsymbol{I} - \boldsymbol{uu}) \cdot \boldsymbol{\kappa} \cdot \boldsymbol{u} dt,$$ [6.47]

which implies that the evolution of the tube is controlled by the strain of the fluid through the velocity gradient $\boldsymbol{\kappa}$. The second equation

$$ds = \sqrt{\frac{2}{\pi^2 \tau_d}} dW,$$ [6.48]

controls the evolution of the chain segment escaping from the tube, and is purely diffusive. $\boldsymbol{W}$ represents a Wiener process.

6.5.2. *Integration of the model*

6.5.2.1. *Functional approximation*

The variational form of the macroscopic description of the flow is given by a variational principle of the Hellinger–Reissner type, as seen in equation [6.30]. This form is estimated by choosing the most adapted interpolation diagrams to the fields of velocity and pressure. It is well known that some approximations do not give stable approximations, but the LBB condition [BAB 73] ensures the stability of the approximation.

Here, a Sibson–Thiessen mixed approach was employed for the fields of velocity and pressure. Even if it is known that this type of approximation can be unstable (it does not verify LBB condition in some precise cases, [GON 04b]), the results obtained are generally stable. Moreover, the approach has a great effectiveness in terms of calculation cost. Other approximations are also possible within a NEM framework, for example, in [GON 04b] an enriched formulation that verifies the LBB condition was proposed. The price to be paid is an additional degree of freedom for each node. More recently, natural element type approximations with conditions of reproducibility [GON 07b] have been proposed. The two approaches give stable and precise results.

6.5.2.2. *Discretization of the model*

Once the interpolation diagram is chosen, the variational form given by equations [6.27]–[6.29] is approximated according to the Bubnov–Galerkin approach:

$$v^h(x) = \sum_{I=1}^{n} \phi_I(x) v_I, \tag{6.49}$$

$$p^h(x) = \sum_{I=1}^{n} \psi_I(x) p_I, \tag{6.50}$$

where n represents the number of nodes considered in the approximation (natural neighbors of point x). Functions $\psi_I(x)$ and $\phi_I(x)$ represent certain forms of interpolation based on the natural neighbors (Thiessen and Sibson, respectively, in the frame of this work). This leads to the system of algebraic equations according to:

$$\begin{pmatrix} \overline{K} & G \\ G^T & M \end{pmatrix} \begin{pmatrix} v \\ p \end{pmatrix} = \begin{pmatrix} f \\ 0 \end{pmatrix}, \tag{6.51}$$

where

$$\overline{K}_{IJ} = \int_\Omega B_I^T \overline{C} B_J \mathrm{d}\Omega \tag{6.52}$$

$$G_{IJ} = -\int_\Omega \tilde{B}_I^T \psi_J \mathrm{d}\Omega \tag{6.53}$$

$$M_{IJ} = -\frac{1}{\lambda} \int_\Omega \psi_I \psi_J \mathrm{d}\Omega \tag{6.54}$$

$$\boldsymbol{f}_I = \int_{\Gamma_t} \phi_I \bar{t} \mathrm{d}\Gamma \qquad [6.55]$$

and

$$\tilde{\boldsymbol{B}}_I = [\phi_{I,1}(\boldsymbol{x}) \quad \phi_{I,2}(\boldsymbol{x})] \qquad [6.56]$$

$$\boldsymbol{B}_I = \begin{pmatrix} \phi_{I,1}(\boldsymbol{x}) & 0 \\ 0 & \phi_{I,2}(\boldsymbol{x}) \\ \phi_{I,2}(\boldsymbol{x}) & \phi_{I,1}(\boldsymbol{x}) \end{pmatrix} \qquad [6.57]$$

$$\overline{C}_{IJKL} = \eta(\delta_{IK}\delta_{JL} + \delta_{IL}\delta_{JK}). \qquad [6.58]$$

It should be noted that, for a completely incompressible case, $M = 0$.

6.5.2.3. *Resolution algorithm*

The method suggested modifies the CONNFFESSIT method. This algorithm, proposed in [LAS 93] consists of the following stages:

– Stage 1: to solve the *macro* equations via the finite element method to obtain the velocity field of fluid v.

– Stage 2: to calculate the trajectory of each molecule by using v.

– Stage 3: to integrate the stochastic differential equation which describes the evolution of the microscopic configuration at the *micro* level.

– Stage 4: to update the viscoelastic stresses.

A sufficient number of molecules must be placed in the geometry of the flow and transported with it. Keunings [KEU 04] cites some associated difficulties, such as tracking the molecules. Schemes which are too simplified usually give bad results. Finally, the number of molecules in each finite element must be sufficiently high (in general several thousand molecules) to avoid statistical noise.

The proposed algorithm takes the following form:

– Stage 1: to solve the *macro* equations via the NEM and obtain the field velocity v.

– Stage 2: to update positions directly, where the molecular trajectory is the node trajectory.

– Stage 3: to integrate the stochastic differential equation for each node.

– Stage 4: to update the viscoelastic stress tensor by doing an average over the configuration space.

In this algorithm, tracking the molecules is not necessary as they are attached to the nodes. As such, no element runs out of its molecules. The price to pay is obtaining the nodal gradient velocities, to carry out stage three. This is not necessary if the method of nodal integration presented in Chapter 3 is employed.

6.5.3. *Some results*

6.5.3.1. *Startup of a simple shear flow of an FENE fluid model*

To validate the suggested approach, here we reproduce the results obtained by traditional techniques.

We consider a simple shear flow defined by $v_x = \dot{\gamma}y$, $v_y = v_z = 0$.

The results obtained in [HER 97] by using the CONNFFESSIT approach for the FENE model were compared with those obtained by using the NEM. A cloud of 525 nodes was used. A homogeneous velocity gradient was imposed on the integration points. The obtained stresses are analyzed below.

The results concerning the evolution of viscosity and the normal stress differences are shown in Figure 6.12 for an arbitrary position in the flow. These evolutions are identical for all the nodes considered. In all the cases, $\lambda_H \cdot \dot{\gamma} = 100.0$ and $b = 20.0$.

Our results are in general good accordance with those of Ottinger. For a long time we have detected some statistical noise, induced by the low number of molecules attached to each node (1000). In what follows we present some more complex flows.

Our results comply with the results obtained by Herrchen and Ötinger [HER 97]. For a long time, we recorded the statistical noise induced by the greatly reduced number of considered molecules (1,000 per node). More complex flows are discussed in the following section.

6.5.3.2. *Flow from an extrusion die and FENE-fluid*

The initial geometry is shown in Figure 6.13.

The fluid flow is forced by inserting nodes of an imposed velocity while following a parabolic law. The maximum value of the velocity is 0.1 m/s. The nodes contain molecules presumed with the state of equilibrium, ψ_{eq}. In this way, the number of nodes of the model increases with time. The time steps were fixed at 0.005 s for the kinematics of the flow as well as for the Euler–Maruyama diagram. Other strategies can also be considered. For example, it is possible to use an algorithm in updating the position of the nodes only after a certain number of time steps of the Euler–Maruyama algorithm. It can be concluded that the Euler–Maruyama diagram requires the use of small time steps compared with that used to update the position of the nodes to produce a correct solution.

Figure 6.14 represents the velocity field for the initial moment and for an intermediate moment of the simulation. The clearly elastic behavior of the fluid is visible.

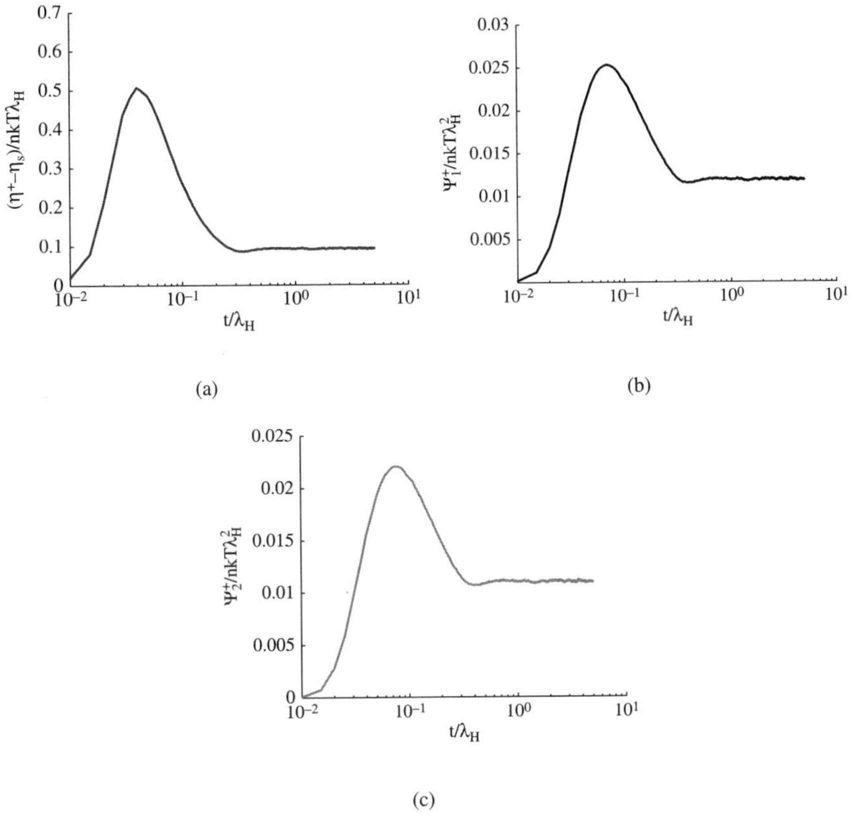

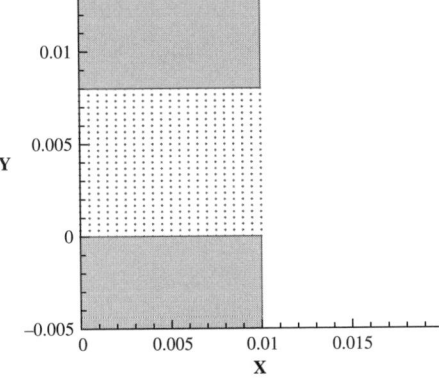

Figure 6.12. *Evolution of viscosity (a), first stress normal difference (b) and second difference in normal stress (c) for an FENE model in a simple shear flow*

Figure 6.13. *Geometry and initial configuration of the nodes in the die*

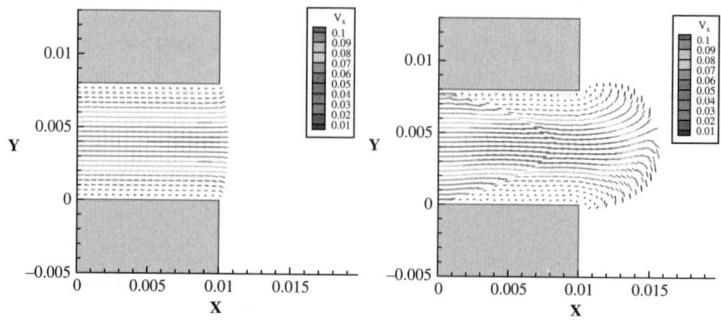

Figure 6.14. *Velocity field at the exit of a die (FENE-fluid) for two different time steps*

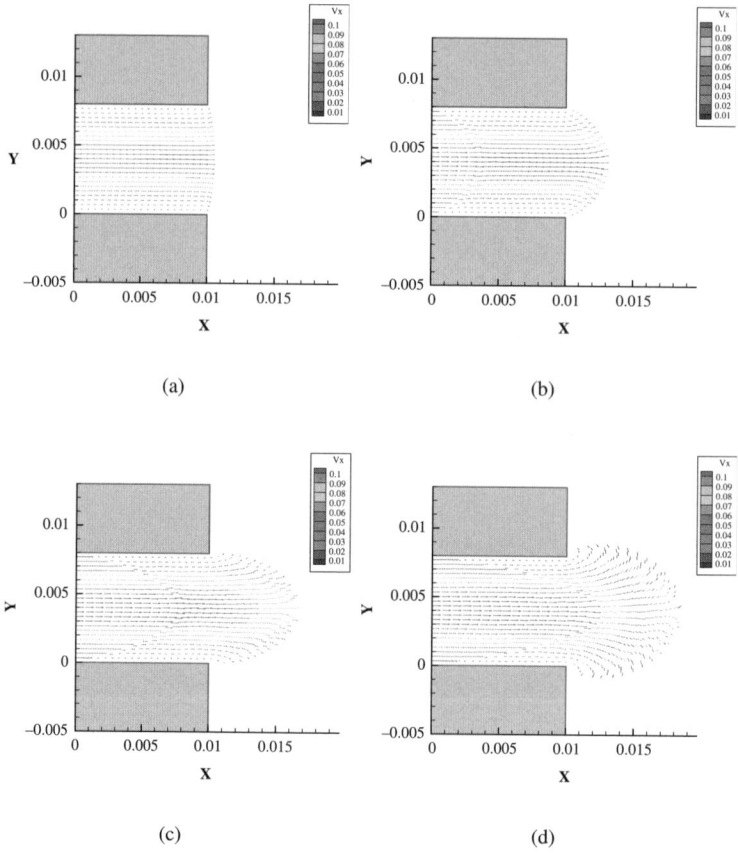

(a)   (b)

(c)   (d)

Figure 6.15. *Evolution of the velocity field at the exit of the die for one polymer in reptation*

6.5.3.3. *Startup of a simple shear flow of a Doi–Edwards fluid model*

The employed geometry is identical to that found in section 6.5.3.2. The flow simulation of a polymer at the exit from a die uses the behavior of the Doi–Edwards type presented in the preceding sections. Time steps of 0.007 s have been employed for this simulation. The number of molecules attached to each node is 100.

Figure 6.15 represents the velocity fields in the flow at four different moments. The elastic effects are still visible in this case. We observe a loss of symmetry after the exit of the die due to the stochastic noise. In spite of the small number of molecules used in the simulation, the stochastic noise remains rather weak.

Chapter 7

Conclusion

This study assesses current developments in the natural element method (NEM) and in particular its extensions, constrained NEM (CNEM) and α-NEM. With the properties of the NEM interpolation functions being very close to those of the finite elements shape functions (local functions in space, partition of unity, linear consistence, and so on), a number of developments presented for the NEM have a neighboring anterior version in the finite elements context (*octree approach*, error indicator, mixed approaches, and so on). This shows that it is completely legitimate to present this new approach as a plausible alternative to the finite elements.

However, the NEM approach offers a notable advantage. It proposes an interpolation that is constructed in a systematic way for a given cloud of nodes. The quality of the obtained interpolation depends on the selected nodal density and not on the relative position of the nodes or some geometrical property of the subjacent elements. For problems evoking great strains, the NEM approach makes it possible to keep the initial nodes throughout the simulation, which is not possible with the finite element approaches because of the geometrical qualities required by the elements that implies frequent remeshing. In particular, this specificity allows the Lagrangian approach where only Eulerian approaches are usually used. Today, this characteristic makes it possible to deal with problems that are not easily accessible with other approaches.

A privileged, and considered difficult, domain of the NEM application is the simulation of material forming processes. Various examples presented in this book (injection, extrusion, shearing, stir friction welding, and so on) show the potential of the approach in this domain and validate the aptitude of NEM to address industrially interesting cases. It is, however, worth noting that all the simulations presented have been carried out using research type software developed in various laboratories. The NEM approach is largely used today, and we look forward to it gaining the attention of developers of commercial computer codes very soon to make it more industrially accessible.

Bibliography

[ALF 06a] ALFARO I., BEL D., CUETO E., DOBLARÉ M., CHINESTA F., "Three-dimensional simulation of aluminium extrusion by the alpha-shape based natural element method", *Computer Methods in Applied Mechanics and Engineering*, vol. 195, no. 33–36, p. 4269–4286, 2006.

[ALF 06b] ALFARO I., YVONNET J., CUETO E., CHINESTA F., DOBLARÉ M., "Meshless methods with application to metal forming", *Computer Methods in Applied Mechanics and Engineering*, vol. 195, no. 48–49, p. 6661–6675, 2006.

[ALF 07] ALFARO I., YVONNET J., CHINESTA F., CUETO E., "A study on the performance of natural neighbour-based Galerkin methods", *International Journal of Numerical Methods Engineering*, vol. 7, no. 12, p. 1436–1465, 2007.

[AMM 06a] AMMAR A., MOKDAD B., CHINESTA F., KEUNINGS R., "A new family of solvers for some classes of multidimensional partial differential equations encountered in kinetic theory modeling of complex fluids", *Journal of Non-Newtonian Fluid Mechanics*, vol. 139, p. 153–176, 2006.

[AMM 06b] AMMAR A., RYCKELYNCK D., CHINESTA F., KEUNINGS R., "On the reduction of kinetic theory models related to finitely extensible dumbbells", *Journal of Non-Newtonian Fluid Mechanics*, vol. 134, p. 136–147, 2006.

[AMM 07] AMMAR A., MOKDAD B., CHINESTA F., KEUNINGS R., "A new family of solvers for some classes of multidimensional partial differential equations encountered in kinetic theory modeling of complex fluids. Part II: Transient simulation using space-time separated representation", *Journal of Non-Newtonian Fluid Mechanics*, vol. 144, p. 98–121, 2007.

[ARN 84] ARNOLD D., BREZZI F., FORTIN M., "A stable finite element for the Stokes equations", *Calcolo*, vol. 21, p. 337–344, 1984.

[ATL 99] ATLURI S., KIM H.G., CHO J.Y., "A critical assessment of the truly meshless local Petrov-Galerkin and local boundary integral equation methods", *Computational Mechanics*, vol. 24, p. 348–372, 1999.

[ATL 00] ATLURI S., ZHU T., "New concepts in meshless methods", *International Journal for Numerical Methods in Engineering*, vol. 47, p. 537–556, 2000.

[ATZ 95] ATZEMA E., HUÉTINK J., "Finite element analysis of forward/backward extrusion using ALE techniques", SHEN S., DAWSON P., (Eds.), *Proc. Numiform 95 Conference*, Ithaca, p. 383–388, 1995.

[BAA 98] BAAIJENS F., An iterative solver for the DEVSS/DG method with application to smooth and non-smooth flows of the upper convected Maxwell fluid", *Journal of Non-Newtonian Fluid Mechanics*, vol. 75, p. 119–138, 1998.

[BAB 73] BABUŠKA I., "The finite element method with Lagrange multipliers", *Numerische Mathematik*, vol. 20, p. 179–192, 1973.

[BAB 79] BABUŠKA I., RHEINBOLDT W., "Adaptive approaches and reliability estimations in finite element analysis", *Computer Methods in Applied Mechanics and Engineering*, vol. 17/18, p. 519–540, 1979.

[BAB 96] BABUŠKA I., MELENK J.M., "The partition of unity finite element method: Basic theory and applications", *Computer Methods in Applied Mechanics and Engineering*, vol. 4, p. 289–314, 1996.

[BAB 97] BABUŠKA I., MELENK J.M., "The partition of unity method", *International Journal for Numerical Methods in Engineering*, vol. 40, p. 727–758, 1997.

[BAB 03] BABUŠKA I., BANERJEE U., OSBORN J.E., "Survey of meshless and generalized finite element methods: A unified approach", *Acta Numerica*, vol. 12, p. 1–125, 2003.

[BAN 05] BANERJEE B., Taylor Impact Tests: Detailed Report, Report no. C-SAFE-CD-IR-05-001. Department of Mechanical Engineering, University of Utah, Salt Lake City, November 2005.

[BAR 93] BARRY J., AN W.C., "Duality of constrained Voronoï diagrams and delaunay triangulations", *Algorithmica*, vol. 9, p. 142–155, 1993.

[BAT 96] BATHE K.J., *Finite Element Procedures*, Prentice-Hall, Englewood Cliffs, 1996.

[BEL 94] BELYTSCHKO T., LU Y.Y., GU L., "Element-free Galerkin methods", *International Journal for Numerical Methods in Engineering*, vol. 37, p. 229–256, 1994.

[BEL 97] BELIKOV V.V., IVANOV V.D., KONTOROVICH V.K., KORYTNIK S.A., SEMENOV A.Y., "The non-Sibsonian interpolation: A new method of interpolation of the values of a function on an arbitrary set of points", *Computational Mathematics and Mathematical Physics*, vol. 37, no. 1, p. 9–15, 1997.

[BEL 98a] BELYTSCHKO T., KRONGAUZ Y., ORGAN D., FLEMING M., KRYSL P., "Meshless methods: An overview and recent developments", *Computer Methods in Applied Mechanics and Engineering*, vol. 139, p. 3–47, 1998.

[BEL 98b] BELYTSCHKO T., LIU W.K., SINGER M., "On adaptivity and error criteria for meshless methods", *Advances in Adaptive Computational Methods in Mechanics*, vol. 47, p. 217–228, 1998.

[BEL 03] BELYTSCHKO T., PARIMI C., MOËS N., SUKUMAR N., USUI S., "Structured extended finite element methods for solids defined by implicit surfaces", *International Journal for Numerical Methods in Engineering*, vol. 56, p. 609–635, 2003.

[BER 94] BERTIN E., Diagrammes de Voronoï 2D et 3D : applications en analyse d'images, PhD Thesis, Joseph Fourier University - Grenoble 1, TIMC - Institut IMAG, January 1994.

[BOI 02] BOISSONNAT J.-D., CAZALS F., "Smooth surface reconstruction via natural neighbour interpolation of distance functions", *Computational Geometry*, vol. 22, p. 185–203, 2002.

[BON 99] BONET J., LOK T.-S.L., "Variational and momentum preservation aspects of smooth particle hydrodynamic formulations", *Computer Methods in Applied Mechanics and Engineering*, vol. 180, p. 97–115, 1999.

[BON 00] BONET J., KULASEGARAM S., "Correction and stabilization of smooth particle hydrodynamics methods with applications in metal forming simulations", *International Journal for Numerical Methods in Engineering*, vol. 47, p. 1189–1214, 2000.

[BON 06] BONITO A., PICASSO M., LASO M., "Numerical simulation of 3D viscoelastic flows with free surfaces", *Journal of Computational Physics*, vol. 215, p. 691–716, 2006.

[BOW 81] BOWYER A., "Computing Dirichlet tesselations", *The Computer Journal*, vol. 24, no. 2, p. 162–166, 1981.

[BRA 83] BRANDES E., *Smithells Metals Reference Book*, Butterworths and Co. Ltd, London, 1983.

[BRA 95a] BRAUN J., SAMBRIDGE M., "A numerical method for solving partial differential equations on highly irregular evolving grids", *Nature*, vol. 376, p. 655–660, 1995.

[BRA 95b] BRAUN J., SAMBRIDGE M., MCQUEEN H., "Geophysical parametrization and interpolation of irregular data using natural neighbours", *Geophysical Journal International*, vol. 122, p. 837–857, 1995.

[BRE 74] BREZZI F., "On the existence, uniqueness and approximation of saddle-point problems arising from Lagrange multipliers", *Revue Française d'Automatique Informatique Recherche Opérationelle, Analyse Numérique*, vol. 8, p. 129–151, 1974.

[BUE 00] BÜELER B., ENGE A., FUKUDA K., "Exact volume computation for polytopes: A practical study", *Polytopes - Combinatorics and Computation*, vol. 29, p. 131–154, 2000.

[BUF 06a] BUFFA G., HU J., SHIVPURI R., FRATINI L., "A continuum based FEM model for friction stir welding-model development", *Materials Science and Engineering A*, vol. 419, p. 389–396, 2006.

[BUF 06b] BUFFA G., HU J., SHIVPURI R., FRATINI L., "Design of the friction stir welding tool using the continuum based FEM model", *Materials Science and Engineering A*, vol. 419, p. 381–388, 2006.

[CAI 05] CAI Y., ZHU H., "A local search algorithm for natural neighbours in the natural element method", *International Journal of Solids and Structures*, vol. 42, p. 6059–6070, 2005.

[CAR 59] CARSLAW H., JAEGER J., *Conduction of Heat in Solids*, Clarendon Press, Oxford, 1959.

[CES 06] CESCOTTO S., LI X., A natural neighbour method for linear elastic problems based on Fraeijs de Veubeke variational principle, Report AG2006-09/01, University of Liège, ArGEnCo Department, 2006.

[CES 07] CESCOTTO S., LI X., "A natural neighbour method for linear elastic problems based on Fraeijs de Veubeke variational principle", *International Journal for Numerical Methods in Engineering*, vol. 71, p. 1081–1101, 2007.

[CES 08] CESCOTTO S., LI X., DUCHENE L., "A natural neighbor method based on Fraeijs de Veubeke variational principle for elastoplasticity and fracture mechanics", *Journal of Engineering Materials and Technology*, vol. 130, 2008.

[CHA 93] CHAPELLE D., BATHE K.J., "The inf-sup test", *Computers and Structures*, vol. 47, no. 4/5, p. 537–545, 1993.

[CHE 01] CHEN J.-S., WU C.-T., YOON S., YOU Y., "A stabilized conforming nodal integration for Galerkin mesh-free methods", *International Journal for Numerical Methods in Engineering*, vol. 50, p. 435–466, 2001.

[CHE 06] CHENOT J.-L., MASSONI E., "Finite element modelling and control of new metal forming processes", *International Journal of Machine Tools and Manufacture*, vol. 46, p. 1194–1200, 2006.

[CHI 05] CHINESTA F., CUETO E., RYCKELYNCK D., AMMAR A., "α-NEM and model reduction: Two new and powerful numerical techniques to describe flows involving short fibers suspensions", *Revue Européenne des Elément Finis*, vol. 14, no. 6–7, p. 903–923, 2005.

[CHU 98] CHUNG H.-J., BELYTSCHKO T., "An error estimate in the EFG method", *Computational Mechanics*, vol. 21, p. 91–100, 1998.

[CUE 00] CUETO E., DOBLARÉ M., GRACIA L., "Imposing essential boundary conditions in the natural element method by means of density-scaled α-shapes", *International Journal for Numerical Methods in Engineering*, vol. 49, no. 4, p. 519–546, 2000.

[CUE 01] CUETO E., MARTÍNEZ M.A., DOBLARÉ M., "El método de los Elementos Naturales en Elasticidad compresible y cuasi-incompresible", *Boletín Técnico del Instituto de Materiales y Modelos Estructurales*, Central University of Venezuela (in Spanish), vol. 39, 2001.

[CUE 02] CUETO E., CALVO B., DOBLARÉ M., "Modeling three-dimensional piece-wise homogeneous domains using the α-shape based natural element method", *International Journal for Numerical Methods in Engineering*, vol. 54, p. 871–897, 2002.

[CUE 03] CUETO E., SUKUMAR N., CALVO B., MARTÍNEZ M.A., CEGOÑINO J., DOBLARÉ M., "Overview and recent advances in natural neighbour Galerkin methods", *Archives of Computational Methods in Engineering*, vol. 10, no. 4, p. 307–384, 2003.

[DE 01] DE S., BATHE K.J., "Towards an efficient meshless computational technique: The method of finite spheres", *Engineering Computations*, vol. 18, p. 170–192, 2001.

[DEL 34] DELAUNAY B., "Sur la Sphère Vide. A la memoire de Georges Voronoï", *Izvestia Akademii Nauk SSSR, Otdelenie Matematicheskii i Estestvennyka Nauk*, vol. 7, p. 793–800, 1934.

[DIR 50] DIRICHLET G.L., "Über die Reduktion der positiven quadratischen Formen mit drei unbestimmten ganzen Zahlen ", *Journal für die Reine und Angewandte Mathematik*, vol. 40, p. 209–227, 1850.

[DOL 99] DOLBOW J., BELYTSCHKO T., "Numerical integration of the Galerkin weak form in meshfree methods", *Computational Mechanics*, vol. 23, p. 219–230, 1999.

[DUA 96] DUARTE C.A.M., ODEN J.T., "An h-p adaptive method using clouds", *Computer Methods in Applied Mechanics and Engineering*, vol. 139, p. 237–262, 1996.

[DUA 00] DUARTE C.A.M., BABUŠKA I., ODEN J.T., "Generalized finite element methods for three-dimensional structural mechanics problems", *Computers and Structures*, vol. 77, no. 2, p. 215–232, 2000.

[EDE 90] EDELSBRUNNER H., MÜCKE E., "Simulation of simplicity: A technique to cope with degenerate cases in geometric algorithms", *ACM Transactions on Graphics*, vol. 9, no. 1, p. 66–104, 1990.

[EDE 94] EDELSBRUNNER H., MÜCKE E., "Three dimensional alpha shapes", *ACM Transactions on Graphics*, vol. 13, p. 43–72, 1994.

[ENG 03] ENGE A., BENNO, FUKUDA K., VINCI Version 1.0.5, Computing Volumes of Convex Polytopes, July 2003.

[FAR 02] FARIN G., *Curves and Surfaces for CAGD*, Morgan Kaufmann, San Francisco, CA, 2002.

[FAR 03] FARIN G., Splines over iterated Voronoï diagrams,   Report no., Arizona State University, 2003.

[FEL 00] FELIPPA C.A., "On the original publication of the general canonical functional of linear elasticity", *Journal of Applied Mechanics*, vol. 67, p. 217–219, 2000.

[FER 04] FERNÁNDEZ-MÉNDEZ S., HUERTA A., BELYTSCHKO T., "Meshless methods", *Encyclopaedia of Computational Mechanics*, on-line edition, DOI: 10.1002/0470091355, 2004.

[FRI 73] FRIED I., "Influence of Poisson's ration on the condition of the finite element stiffness matrix", *International Journal of Solids and Structures*, vol. 10, p. 323–329, 1973.

[GÉR 96] GÉRADIN M., RIXEN D., *Théorie des vibrations : application à la dynamique des structures*, Masson, 1996.

[GIG 02] GIGRAS P.G., KHOMAMI B., "Adaptive configuration fields: A new multiscale simulation technique for reptation-based models with a stochastic strain measure and local variations of life span distribution", *Journal of Non-Newtonian Fluid Mechanics*, vol. 108, no. 1-3, p. 99–122, 2002.

[GON 04a] GONZALEZ D., CUETO E., MARTINEZ M.A., DOBLARE M., "Numerical integration in natural neighbour Galerkin methods", *International Journal for Numerical Methods in Engineering*, vol. 60, no. 12, p. 2077–2104, 2004.

[GON 04b] GONZÁLEZ D., CUETO E., DOBLARÉ M., "Volumetric locking in natural neighbour Galerkin methods", *International Journal for Numerical Methods in Engineering*, vol. 61, no. 4, p. 611–632, 2004.

[GON 07a] GONZÁLEZ D., CUETO E., CHINESTA F., DOBLARÉ M., "A natural element updated Lagrangian strategy for free-surface fluid dynamics", *Journal of Computational Physics*, vol. 223, no. 1, p. 127–150, 2007.

[GON 07b] GONZÁLEZ D., CUETO E., DOBLARÉ M., "Higher-order natural element methods: Towards an isogeometric meshless method", *International Journal for Numerical Methods in Engineering*, vol. 74, no. 13, p. 1928–1954, 2007.

[GON 08] GONZALEZ D., CUETO E., DOBLARE M., "Higher-order natural element methods: Towards an isogeometric meshless method", *International Journal for Numerical Method in Engineering*, vol. 74, no. 13, p. 1928–1954, 2008.

[GOS 96] GOSZ J., LIU W.K., "Admissible approximations for essential boundary conditions in the reproducing kernel particle method", *Computational Mechanics*, vol. 19, p. 120–135, 1996.

[GRE 78] GREEN P.J., SIBSON R., "Computing dirichlet tesselations in the plane", *The Computer Journal*, vol. 21, p. 168–173, 1978.

[HER 97] HERRCHEN M., ÖTTINGER H.C., "A detailed comparison of various FENE models", *Journal Non-Newtonian Fluid Mechanics*, vol. 68, p. 17–42, 1997.

[HIY 99] HIYOSHI H., SUGIHARA K., "Two generalizations of an interpolant based on Voronoï diagrams", *International Journal of Shape Modeling*, vol. 5, no. 2, p. 219–231, 1999.

[HU 55] HU H.C., "On some variational principles in the theory of elasticity and the theory of plasticity", *Scientia Sinica*, vol. 4, p. 33–54, 1955.

[HUG 87] HUGHES T.J.R., FERENCZ R.M., HALLQUIST J.O., "Large-scale vectorized implicit calculations in solid mechanics on a Cray X-MP/48 utilizing EBE preconditioned conjugate gradients", *Computer Methods in Applied Mechanics and Engineering*, vol. 61, p. 215–248, 1987.

[ILL 08] ILLOUL L., Mise en œuvre de la méthode des éléments naturels contrainte en 3D – Application au cisaillage adiabatique, PhD Thesis, Ecole Nationale Supérieure d'Arts et Métiers, 2008.

[IRO 72] IRONS B., RAZZAQUE A., "Experience with the patch test for convergence of finite elements", AZIZ A.K., (Ed.), *The Mathematical Foundations of the Finite Element Method with Applications to Partial Differential Equations*, Academic Press, New York, 1972.

[JI 02] JI H., CHOPP D., DOLBOW J., "A hybrid finite element/level set method for modeling phase transformations", *International Journal for Numerical Methods*, vol. 54, p. 1209–1233, 2002.

[KAG 03] KAGAN P., FISCHER A., BAR-YOSEPH P.Z., "Mechanically based models: Adaptive refinement of B-spline finite element", *International Journal for Numerical Methods in Engineering*, vol. 57, p. 1145–1175, 2003.

[KEU 04] KEUNINGS R., "Micro-macro methods for the multiscale simulation of viscoelastic flow using molecular models of kinetic theory", BINDING D., WALTERS K. (Eds.), *Rheology Reviews 2004*, British Society of Rheology, Aberystwyth, Wales, p. 67-98, 2004.

[KEY 90] KEYAK J.H., MEAGHER J.M., "Automated three dimensional finite element modelling of bone: A new method", *Journal of Biomedical Engineering*, vol. 12, p. 389–397, 1990.

[KLA 00] KLAAS O., SHEPHARD M.S., "Authomatic generation of octree based three-dimensional discretizations for partition of unity methods", *Computational Mechanics*, vol. 25, p. 296–304, 2000.

[KRO 96] KRONGAUZ Y., Application of meshless methods to solid mechanics, PhD Thesis, Northwestern University, Evanston, IL, 1996.

[KRY 03] KRYSL P., GRINSPUN E., SCHRÖDER P., "Natural hierarchical refinement for finite element methods", *International Journal for Numerical Methods in Engineering*, vol. 56, p. 1109–1124, 2003.

[LAD 98] LADEVÈZE P., *Non Linear Computational Structural Mechanics*, Springer, New York, 1998.

[LAD 01] LADEVÈZE P., PELLE J.-P., *La Maîtrise du calcul en mécanique linéaire et non-linéaire*, Hermes/Lavoisier, 2001.

[LAN 00] VAN DE LANGKRUIS J., LOF J., KOOL W.H., VAN DER ZWAAG S., HUÉTINK J., "Comparison of experimental AA6063 extrusion trials to 3D numerical simulations, using a general solute-dependent constitutive model", *Computational Materials Science*, vol. 81, p. 381–392, 2000.

[LAS 93] LASO M., ÖTTINGER H.C., "Calculation of viscoelastic flow using molecular models: the CONNFFESSIT approach", *Journal of Non-Newtonian Fluid Mechanics*, vol. 47, p. 1–20, 1993.

[LAW 91] LAWRENCE J., "Polytope volume computation", *Mathematics of Computation*, vol. 57, p. 259–271, 1991.

[LEE 03a] LEE C.K., ZHOU C.E., "On error estimation and adaptive refinement for element free Galerkin method: Part I: Stress recovery and a posteriori error estimation", *Computer and Structures*, vol. 82, p. 413–428, 2003.

[LEE 03b] LEE C.K., ZHOU C.E., "On error estimation and adaptive refinement for element free Galerkin method: Part II: Adaptive refinement", *Computer and Structures*, vol. 82, p. 429–443, 2003.

[LEW 97] LEWIS R.W., NAVTI S.E., TAYLOR C., "A mixed Lagrangian-Eulerian approach to modelling fluid flow during mould filling", *International Journal for Numerical Methods in Engineering*, vol. 25, p. 931–952, 1997.

[LI 07] LI X., CESCOTTO S., ROSSI B., A natural neighbour method for materially non linear problems based on Fraeijs de Veubeke variational principle, Report AG2007-03/01, University of Liège, ArGEnCo Department, 2007.

[LI 08] LI X., CESCOTTO S., ROSSI B., "A natural neighbour method for materially non linear problems based on Fraeijs de Veubeke variational principle", *Acta Mechanica Sinica*, vol. 25, p. 83–93, 2009.

[LIU 95] LIU W.K., JUN S., LI S., ADEE J., BELYTSCHKO T., "Reproducing kernel particle methods", *International Journal for Numerical Methods in Engineering*, vol. 38, p. 1655–1679, 1995.

[LIU 97] LIU W.K., URAS R.A., CHEN Y., "Enrichment of the finite element method with the reproducing kernel particle method", *Journal of Applied Mechanics, ASME*, vol. 64, p. 861–870, 1997.

[LOF 99] LOF J., HUÉTINK J., "Numerical simulation of the aluminium extrusion process", COVAS J. (Ed.), *Proc. 2nd ESAFORM Conference on Material Forming*, p. 29–32, 1999.

[LOF 00] LOF J., Developments infinite element simulations of aluminium extrusion, PhD Thesis, Department of Mechanical Engineering, Applied Mechanics Section, University of Twente, The Netherlands, 2000.

[LOF 01] LOF J., "Elasto-viscoplastic FEM simulation of the aluminium flow in the bearing area for extrusion of thin-walled sections", *Journal of Materials Processing Technology*, vol. 114, p. 174–183, 2001.

[LOF 02] LOF J., BLOKHUIS Y., "FEM simulations of the extrusion of complex thin-walled aluminium sections", *Journal of Materials Processing Technology*, vol. 122, p. 344–354, 2002.

[LOR 87] LORENSEN W.E., CLINE H.E., "Marching cubes: A high resolution 3D surface reconstruction algorithm", *Computer Graphics*, vol. 21, p. 163–169, 1987.

[LU 02] LU H., CHEN J., "Adaptive meshfree particle method", *Lecture Notes in Computational Science and Engineering*, vol. 26, p. 251–267, 2002.

[LYN 81] LYNCH D., O'NEILL K., "Continuously deforming finite elements for the solution of parabolic problems, with and without phase change", *International Journal for Numerical Methods in Engineering*, vol. 17, p. 81–96, 1981.

[MAC 03] MACRI M., DE S., SHEPHARD M.S., "Hierarchical tree-based discretization for the method of finite spheres", *Computers and Structures*, vol. 81, p. 789–803, 2003.

[MAR 52] MARTIN J.C., MOYCE W.J., "Part IV: An experimental study of the collapse of liquid columns on a rigid horizontal plane", *Philosophical Transactions of the Royal Society London*, vol. 244, p. 312, 1952.

[MAR 02] MARTÍNEZ M.A., CUETO E., DOBLARÉ M., CHINESTA F., "Natural element meshless simulation of injection processes involving short fiber suspensions", *Internatonal Journal of Forming Processes*, vol. 4, no. 3-4, 2002.

[MAR 03] MARTÍNEZ M.A., CUETO E., DOBLARÉ M., CHINESTA F., "Fixed mesh and meshfree techniques in the numerical simulation of injection processes involving short fiber suspensions", *Journal of Non-Newtonian Fluid Mechanics*, vol. 115, p. 51–78, 2003.

[MAR 04] MARTÍNEZ M.A., CUETO E., ALFARO I., DOBLARE M., CHINESTA F., "Updated Lagrangian free surface flow simulations with natural neighbour Galerkin methods", *International Journal for Numerical Methods in Engineering*, vol. 60, no. 13, p. 2105–2129, 2004.

[MEA 82] MEAGHER D., "Geometric modeling using octree encoding", *Computer Graphics and Image Processing*, vol. 19, p. 129–147, 1982.

[MER 02] MERLE R., DOLBOW J., "Solving thermal and phase change with the extended finite element method", *Computational Mechanics*, vol. 28, no. 5, p. 339–350, 2002.

[MOO 95] MOOI H.G., HUÉTINK J., "Simulation of complex aluminium extrusion using an arbitrary Eulerian Lagrangian formulation", SHEN S.F., DAWSON P.R. (Eds.), *Proc. Numiform 95 Conference*, New York, p. 869–874, 1995.

[MUE 93] MUECKE E.P., Shapes and implementations in three-dimensional geometry, PhD Thesis, University of Illinois, Urbana-Champaign, 1993.

[NAG 02] NAGASHIMA T., ISHIHARA Y., NIIYIMA K., MAKINOUCHI A., "Development of stress analysis method based on Voxel-type X-FEM", MANG H.A., RAMMERSTORFER F.G., EBERHARDSTEINER J., (Eds.), *Proc. 5th World Congress on Compotatinal Mechanics, Vienna*, IACM, 2002.

[NAY 92] NAYROLES B., TOUZOT G., VILLON P., "Generalizing the finite element method: Diffuse approximation and diffuse elements", *Computational Mechanics*, vol. 10, p. 307–318, 1992.

[ODE 98] ODEN J.T., DUARTE C.A.M., ZIENKIEWICZ O.C., "A new cloud based hp finite element method", *Computer Methods in Applied Mechanics and Engineering*, vol. 153, p. 117–126, 1998.

[ORG 96] ORGAN D., FLEMING M., TERRY T., BELYTSCHKO T., "Continuous meshless approximations for nonconvex bodies by diffraction and transparency", *Computational Mechanics*, vol. 18, p. 1–11, 1996.

[ÖTT 96] ÖTTINGER H.C., *Stochastic Processes in Polymeric Fluids*, Springer, New York, 1996.

[OWE 02] OWENS R.G., PHILLIPS T.N., *Computational Rheology*, Imperial College Press, London, 2002.

[PER 66] PERZYNA P., "Fundamental problems in visco-plasticity", *Recent Advances in Applied Mechanics*, Academic Press, New York, 1966.

[RAB 03] RABIER S., MEDALE M., "Computation of free surface flows with a projection FEM in a moving mesh framework", *Computer Methods in Applied Mechanics and Engineering*, vol. 192, no. 41–42, 2003.

[RAM 87] RAMASWAMY B., KAWAHARA M., "Lagangian finite element analysis applied to viscous free surface fluid flow", *International Journal for Numerical Methods in Engineering*, vol. 7, p. 953–984, 1987.

[RAN 04] RANC N., Etude des champs de température et de déformation dans les matériaux métalliques sollicités à grande vitesse de déformation, PhD Thesis, University Paris X, Nanterre, December 2004.

[RIE 96] VAN RIETBERGEN B., WEINANS H., HUISKES R., "Computational strategies for iterative solutions of large FEM applications employing Voxel data", *International Journal for Numerical Methods in Engineering*, vol. 39, p. 2473–2767, 1996.

[ROS 07] ROSSI B., HABRAKEN A.M., PASCON F., "On the evaluation of the through thickness residual stresses distribution of cold formed profiles", *Proc. 10th ESAFORM Conference on Material Forming, Zaragoza*, p. 570–577, 2007.

[RVA 95] RVACHEV V.L., SHEIKO T.I., "R-functions in boundary value problems in Mechanics", *Applied Mechanics Reviews*, vol. 48, p. 151–188, 1995.

[RVA 00] RVACHEV V.L., SHEIKO T.I., SHAPIRO V., TSUKANOV I., "On completeness of RFM solution structures", *Computational Mechanics*, vol. 25, p. 305–316, 2000.

[SAM 84] SAMET H., "The quadtree and related hierarchical data structures", *Computing Surveys*, vol. 16, no. 2, p. 187–260, 1984.

[SAM 95] SAMET H., "Spatial data structures", *Modern Database Systems: The Object Model, Interoperability and Beyond*, Addison Wesley/ACM press, Reading, MA, p. 361–385, 1995.

[SAM 96] SAMBRIDGE M., BRAUN J., MCQUEEN H., "New computational methods for natural neighbour interpolation in two and three dimensions", *Computational Techniques and Applications : CTAC95*, p. 685–692, 1996.

[SCH 28] SCHÖNHARDT E., "Über die Zerlegung von Dreieckspolyedern in Tetraeder", *Mathematische Annalen*, vol. 98, p. 309–312, 1928.

[SCH 90] SCHMITT F., BOROUCHAKI H., "Algorithme rapide de maillage de Delaunay dans R*d*", *Journées de Géométrie Algorithmique*, p. 131–133, 1990.

[SEL 72] SELLARS C.M., TEGART W.J.M., "Hot workability", *International Metallurgical Review*, vol. 17, p. 1–24, 1972.

[SHA 99] SHAPIRO V., TSUKANOV I., "Meshfree simulation of deforming domains", *Computer-Aided Design*, vol. 31, p. 459–471, 1999.

[SHE 98] SHEWCHUK J.R., "A condition guaranteeing the existence of higher-dimensional constrained delaunay triangulations", *Proc. 14th Annual Symposium on Computational Geometry*, Minnesota, p. 76–85, 1998.

[SI 05] SI H., GÂRTNER K., "Meshing piecewise linear complexes by constrained delaunay tetrahedralizations", *Proc. 14th International Meshing Roundtable*, San Diego, CA, p. 147–163, 2005.

[SIB 80] SIBSON R., "A vector identity for the dirichlet tesselation", *Mathematical Proc. Cambridge Philosophical Society*, vol. 87, p. 151–155, 1980.

[SIB 81] SIBSON R., "A brief description of natural neighbour interpolation", BARNETT V. (Ed.), *Interpreting Multivariate Data*, John Wiley, Chichester, p. 21–36, 1981.

[STR 01] STROUBOULIS T., COPPS K., BABUŠKA I., "The generalized finite element method", *Computer Methods in Applied Mechanics and Engineering*, vol. 190, p. 4081–4193, 2001.

[SUK 98a] SUKUMAR N., The natural element method in solid mechanics, PhD Thesis, Northwestern University, Evanston, Illinois, 1998.

[SUK 98b] SUKUMAR N., MORAN B., BELYTSCHKO T., "The natural element method in solid mechanics", *International Journal for Numerical Methods in Engineering*, vol. 43, no. 5, p. 839–887, 1998.

[SUK 01a] SUKUMAR N., CHOPP D.L., MOËS N., BELYTSCHKO T., "Modeling holes and inclusions by level sets in the extended finite element method", *Computer Methods in Applied Mechanics and Engineering*, vol. 190, no. 46–47, p. 6183–6200, 2001.

[SUK 01b] SUKUMAR N., MORAN B., SEMENOV A.Y., BELIKOV V.V., "Natural neighbor Galerkin methods", *International Journal for Numerical Methods in Engineering*, vol. 50, no. 1, p. 1–27, 2001.

[SUK 05] SUKUMAR N., DOLBOW J., DEVAN A., YVONNET J., CHINESTA F., RYCKELYNCK D., LORONG P., ALFARO I., MARTÍNEZ M.A., CUETO E., DOBLARÉ M., "Meshless methods and partition of unity finite elements", *Internatonal Journal of Forming Processes*, vol. 8–4, p. 409–427, 2005.

[THI 11] THIESSEN A.H., Precipitation averages for large areas, Monthly Weather Report, vol. 39, p. 1082–1084, 1911.

[TIM 72] TIMOSHENKO S., GOODIER J.N., *Teoría de la Elasticidad*, Editorial Urmo, 1972.

[TRA 94] TRAVERSONI L., "Natural neighbour finite elements", *International Conference on Hydraulic Engineering Software, Hydrosoft Proceedings*, Computational Mechanics publications, Southampton, p. 291–297, 1994.

[TRU 05] TRUNZLER J., JOYOT P., CHINESTA F., "Discontinuous derivative enrichment in RKPM", *Lecture Notes on Computational Science and Engineering*, vol. 43, Springer, p. 93–108, 2005.

[VEU 51] FRAEIJS DE VEUBEKE B.M., "Diffusion des inconnues hyperstatiques dans les voilures à longerons couplés", *Bulletins du Service Technique de l'Aéronautique*, vol. 24, 1951.

[VOR 08] VORONOÏ G. M., "Nouvelles Applications des Paramètres Continus à la Théorie des Formes Quadratiques. Deuxième Memoire: Recherches sur les parallélloèdres Primitifs", *Journal für die Reine and Angewandte Mathematic*, vol. 134, p. 198–287, 1908.

[WAS 55] WASHIZU K., On the variational principles of elasticity and plasticity, Aeroelastic and Structures Research Laboratory, Technical report 25-18, MIT, Cambridge, 1955.

[WAT 81] WATSON D., "Computing the n-dimensional delaunay tessellation with application to Voronoï polytopes", *The Computer Journal*, vol. 24, no. 2, p. 162–172, 1981.

[WAT 01] WATSON D.F., Compound signed decomposition, the core of natural neighbour interpolation in n-dimensional space (unpublished manuscript), ©2001 D.F. Watson, http://www.lang.org/naturalneighbour.html

[WEL 02] WELLS G.N., SLUYS L.J., DE BORST R., "A p-adaptive scheme for overcoming volumetric locking during plastic flow", *Computer Methods in Applied Mechanics and Engineering*, vol. 191, p. 3153–3164, 2002.

[WIL 65] WILKINSON J.H., *The Algebraic Eigenvalue Problem*, Clarendon Press, Oxford, 1965.

[YER 84] YERRY M.A., SHEPHARD M.S., "Automatic three-dimensional mesh generation by the modified octree technique", *International Journal for Numerical Methods in Engineering*, vol. 20, p. 1965–1990, 1984.

[YOO 04] YOO J., MORAN B., CHEN J.-S., "Stabilized conforming nodal integration in the natural-element method", *International Journal for Numerical Methods in Engineering*, vol. 60, p. 861–890, 2004.

[YOU 03] You Y., Chen J., Lu H., "Filters, reproducing Kernel, and adaptive meshfree method", *Computational Mechanics*, vol. 31, p. 316–326, 2003.

[YVO 04a] Yvonnet J., Nouvelles approches sans maillage basées sur la méthode des éléments naturels pour la simulation numérique des procédés de mise en forme, PhD Thesis, Ecole Nationale Supérieure d'Arts et Métiers, 2004.

[YVO 04b] Yvonnet J., Ryckelynck D., Lorong P., Chinesta F., "A new extension of the natural element method for non-convex and discontinuous problems: The constrained natural element method", *International Journal for Numerical Methods in Enginering*, vol. 60, no. 8, p. 1452–1474, 2004.

[ZHO 03] Zhou J., Li L., Duszczyk J., "3D FEM simulation of the whole cycle of aluminium extrusion throughout the transient state and the steady state using the updated Lagrangian approach", *Journal of Materials Processing Technology*, vol. 134, p. 383–397, 2003.

[ZIE 74] Zienkiewicz O.C., Godbolet P.N., "Flow of plastic and visco-plastic solids with special reference to extrusion and forming processes", *International Journal for Numerical Methods in Engineering*, vol. 8, p. 3–16, 1974.

[ZIE 78a] Zienkiewicz O.C., Oñate E., Heinrich J.C., "Plastic flow in metal forming. (I) Coupled thermal (II) Thin sheet forming", *Applications of Numerical Methods to Forming Processes. AMD-vol 28*, p. 107–120, 1978.

[ZIE 78b] Zienkiewicz O.C., Pain P.C., Oñate E., "Flow of solids during forming and extrusion: Some aspects of numerical solutions", *International Journal of Solids and Structures*, vol. 14, p. 15–38, 1978.

[ZIE 84] Zienkiewicz O.C., "Flow formulation for numerical solution of forming processes", Pittman J.F.T., Zienkiewicz O.C., Wood R.D., Alexander J.M. (Eds.), *Numerical Analysis of Forming Processes*, John Wiley and Sons, Chichester, p. 1–44, 1984.

[ZIE 88] Zienkiewicz O.C., Zhu J.Z., "A simple error estimator and adaptive procedure for practical engineering analysis", *International Journal for Numerical Methods in Engineering*, vol. 24, p. 337–357, 1988.

[ZIE 99] Zienkiewicz O., Boroomand B., Zhu J., "Recovery procedures in error estimation and adaptivity Part I: Adaptivity in linear problems", *Computer Methods in Applied Mechanics Engineering*, vol. 26, p. 111–125, 1999.

Index